This book is to be returned on or before
the last date stamped below.

CHADWICK . T. F.                    540

## WESTON - SUPER - MARE
## COLLEGE LIBRARY

chemistry : a modern introduction
for sixth forms

LIBREX

# CHEMISTRY: A MODERN INTRODUCTION FOR SIXTH FORMS AND COLLEGES

PART 2
## Organic Chemistry

Second edition

T. F. Chadwick, B.Sc., C.Chem., F.R.I.C., Dip. Chem. Eng.
*College of Technology*
*Reading*

*London*
*GEORGE ALLEN & UNWIN LTD*
RUSKIN HOUSE   MUSEUM STREET

FIRST PUBLISHED IN 1971

SECOND IMPRESSION 1974

SECOND EDITION 1977

© *George Allen & Unwin (Publishers) Ltd 1971, 1977*

ISBN 0 04 547003 0

*Printed in Great Britain
in 10 on 11 Point Times Roman
by William Clowes & Sons, Limited
London, Beccles and Colchester*

# PREFACE TO THE SECOND EDITION

In the few years which have elapsed since the publication of the first edition of this book, several important changes relating to the presentation of Organic Chemistry in 'A' level and National Certificate courses have taken place. One of the most important of these changes has been the increasing use of the ASE system of nomenclature and the new recommended names have been used consistently throughout the second edition. Reaction mechanisms, which were introduced tentatively in the first edition, have been carefully revised and extended as the importance and value of a study of this aspect of the subject is now firmly established at this level. Several chapters have been extended in order to bring the subject matter up to date, especially those dealing with biologically important compounds, compounds of industrial importance and the isolation and analysis of organic compounds.

The aim of the book is unchanged from that of the first edition, that is to present in a concise form the factual framework of elementary organic chemistry, reinforced by an introduction to reaction mechanisms which makes for a better understanding of the subject. Again, summary chapters are included together with tables and charts as these are felt to be of value for revision purposes.

T. F. CHADWICK

# CONTENTS

# 1 The General Characteristics of Organic Chemistry

The outstanding property of carbon is its ability to form a vast number of covalent compounds, and the study of these compounds is designated *organic chemistry*.

The chief characteristics of organic chemistry may be summarized as follows:

1 Organic compounds may be simple molecules containing relatively few atoms (for example, methane, $CH_4$), or highly complicated molecules containing many atoms (such as cholesterol, $C_{27}H_{45}O$).

2 Although a large number of atoms are often found in organic molecules, the atoms involved are mainly confined to a small number of different elements. All organic compounds contain carbon, and (with a few exceptions) hydrogen. Oxygen, nitrogen and the halogens are often present, while sulphur and phosphorus appear in a few organic compounds.

3 In organic compounds, carbon shows a covalency of four, oxygen two, the halogens and hydrogen one. A single covalent bond results from the sharing of a pair of electrons between two atoms (see Part 1). This is written

$$H:H \quad \text{or} \quad H - H$$

where the two dots or a dash indicate the shared electron pair.

4 Because large numbers of atoms of the same elements are present in organic compounds, the differences between one organic molecule and another lie in the way in which the various atoms are linked together. Consequently, a knowledge of the *structure* of an organic molecule is important, and it is frequently found that two (or more) molecules may contain the same number of atoms of the same elements and yet have different structures, and, therefore, different properties. This is known as *isomerism* (see Chapter 16).

5 Carbon atoms may be linked to each other by single (one shared electron pair), double (two shared electron pairs) or triple (three shared electron pairs) bonds. These are written as:

$$C:C \quad \text{or} \quad -\overset{\mid}{\underset{\mid}{C}}-\overset{\mid}{\underset{\mid}{C}}-$$

$$C::C \quad \text{or} \quad \overset{}{\underset{}{C}}{=}\overset{}{\underset{}{C}}$$

$$C:::C \quad \text{or} \quad -C{\equiv}C-$$

Carbon can also link to other atoms by means of multiple bonds—with oxygen for example:

$$\overset{}{\underset{}{C}}{=}O$$

6 A succession of carbon atoms may be linked together, and organic compounds are generally divided into three main classes according to the form of the molecule produced by such successive interlinking. The classes are:

(a) *Aliphatic*, or *open chain* compounds, in which the carbon atoms at the ends of the chain are bonded to atoms of a different element (i.e. the chain has a starting point and a terminal point). Typical aliphatic compounds are propane and chloroethanoic acid.

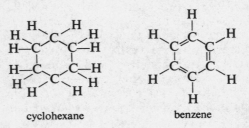

propane        chloroethanoic acid

(b) *Cyclic*, or *closed chain* compounds, in which the carbon end atoms in a chain are linked to form a ring structure. Typical cyclic compounds are cyclohexane and benzene.

cyclohexane        benzene

Compounds containing the six-membered ring structure found in benzene are known as *aromatic* compounds.

(c) *Heterocyclic* compounds, which are those where the ring structure contains an atom of an element other than carbon. Furan and pyridine are typical heterocyclic compounds.

6

```
    H                              H
     \                              |
      C ═══ C                       C
     /       \                     / \
    H         H           H       C   C       H
     \                      \    ‖     ‖    /
      C ═══ C                 C         C
     /  \  /  \             /    \     /    \
    H    O    H            H      C ═ N      H
                                 /      \
                                H        H
```

furan                        pyridine

7 Reaction between covalent molecules involves the breaking and re-making of bonds. For this to take place, a minimum amount of energy (known as the activation energy, see Part 3) must be available. The energy of a system rises as the temperature increases, consequently many organic reactions which take place slowly when cold, speed up as the temperature rises. In general, the speed of a typical organic reaction approximately doubles following a temperature rise of 10 degrees. Quite often, boiling under reflux may be needed to give a reasonable yield of product. (Some organic compounds contain ionic bonds, and reactions involving these ions take place instantaneously in ionizing solvents.)

8 As a result of the random motion of molecules, different bonds may be ruptured on the collision of molecules. This leads to the possible formation of several substances among the products of a reaction. In practice, under the prevailing conditions of the reaction, a large proportion of one substance (called the *major product*) together with smaller quantities of *side products* are formed. Therefore the same reaction may, under different conditions, give rise to different major products. In addition, the major product has to be separated and purified from the collection of substances (both side products and unchanged reactants) present in the reaction mixture. Techniques of product recovery and purification are discussed in the following chapter.

# 2 The Isolation and Analysis of Organic Compounds

Before an organic compound can be fully investigated, it has to be recovered from the reaction mixture used for its preparation, or extraction (in the case of a substance being extracted from a naturally occurring material). After this, the compound must be purified and analysed qualitatively and quantitatively. Briefly, the techniques commonly used are as follows:

## Product Recovery

**Distillation** This method is widely used when the major product is a liquid. If the boiling points of the major product and the other liquids (side products, reactants, solvent, etc.) are widely different, simple distillation is effective. Where the boiling points of the mixture are closer together, *fractional* distillation (carried out in an apparatus which allows a repeated series of simple distillations to be effected within the apparatus) is employed. For liquids of high boiling points, *vacuum* distillation may be used, since the compound may decompose appreciably at the high temperatures otherwise involved. *Steam* distillation is used for liquids which are immiscible in water, yet volatile in steam, while special techniques are employed for liquids that form a constant boiling mixture (azeotrope). More details of distillation techniques are included in Part 3 of this series.

**Precipitation** When the product is a solid held in solution by the solvent, precipitation of the solid product may be brought about by:

(a) acidification of the solution where this is initially alkaline,
(b) adding alkali to an acid solution,
(c) strong cooling of the solution,
(d) diluting the solution with a liquid in which the product does not dissolve (e.g. adding water to an alcoholic solution of a water-insoluble product).

The precipitate is then filtered off or centrifuged and purified.

**Solvent extraction** The most commonly used solvent in this technique is ether, and the method is chiefly used for the extraction of organic substances from aqueous solution. When ether and the aqueous solution are well shaken, it is usually found that the organic material is more soluble in ether, and is extracted into the ether layer. After the ether layer has been separated from the aqueous layer, the ether is removed by distillation or evaporation (away from all naked flames) and the product is purified.

## Purification

Liquids are usually purified by distillation, although fractional crystallization is sometimes used—benzene, for example (m.p. 5·5°C) is purified this way. Solids are purified by *recrystallization*. In this process, the impure solid is dissolved in the minimum quantity of hot solvent, thus forming an almost saturated solution. When this solution (filtered free from any insoluble particles if necessary) is cooled, crystals of purer material are deposited. One or more recrystallizations may be required, depending on the purity of the starting material.

*Sublimation* is an alternative technique for the purification of solids, and this method is sometimes used when a suitable solvent for recrystallization cannot be found.

# ORGANIC ANALYSIS AND STRUCTURAL FORMULAE

Having obtained a purified compound, the following investigations are carried out in order to elucidate the structural formula of the compound:

1 Tests to determine the purity of the compound.
2 The presence of the various elements in the compound is shown by qualitative analysis.
3 The weight of each element present is determined by quantitative analysis.
4 The empirical formula for the compound is calculated.
5 The relative molecular mass is determined and the molecular formula for the compound is deduced.
6 A study of the chemistry of the compound indicates the most probable grouping of the atoms in the molecule, thus leading to a suggested structural formula.

We now discuss each of these stages in more detail.

## Testing the Purity of a Compound

An indication of the purity of an unknown compound is given by various physical tests. For a solid, a melting point determination is usually employed. A small quantity of the compound is introduced into a fine capillary tube which is sealed at its lower end. The capillary tube is attached to a thermometer and the two are heated slowly in a well stirred oil bath or electrically heated block. The temperature at which the sample first begins to melt and that at which melting is complete are taken. This temperature interval is known as the melting range; it is found that pure substances melt very sharply within a narrow melting range. Further recrystallization should show no change in the melting range of a pure compound, but in the case of one which does contain some impurities, further purification leads to a rise in the melting point and a narrowing of the melting range because the presence of impurities almost always lowers the melting point. This fact is made use of in the technique of *mixed melting point* determinations, which is widely used to identify organic compounds. Should a compound under test appear to be X for instance, then a mixture of the sample and X should have the same melting point as pure X if this is in fact the case.

Chromatographic methods can also be used to assess the purity of a compound.

In the case of liquids, determination of the boiling range of a compound can be used to give an indication of purity. Since boiling points vary with pressure it is not as easy to assess the purity of a compound from these measurements as it is in the case of melting point determinations. However, chromatographic methods of analysis are particularly well suited to this type of problem, and detailed information about the purity of a compound can be obtained in this way. The use of chromatographic analysis in relation to gases is equally important.

## Qualitative Analysis

**Carbon and hydrogen**   The presence of these elements is established by oxidation to carbon dioxide and water respectively, using the apparatus shown in Figure 1.

The compound under test is mixed with dry copper (II) oxide and heated in a hard glass test tube. Water vapour produced by the oxidation of hydrogen in the compound is indicated by a blue coloration produced with the anhydrous copper (II) sulphate, while cloudiness appearing in the lime water confirms the presence of carbon dioxide produced by oxidation of carbon in the compound.

**Nitrogen, sulphur, phosphorus and the halogens**   These elements are detected by the *Lassaigne sodium fusion* technique. In this test, a small portion of the organic compound is allowed to react with excess molten sodium in an ignition tube, heating being continued until all

10

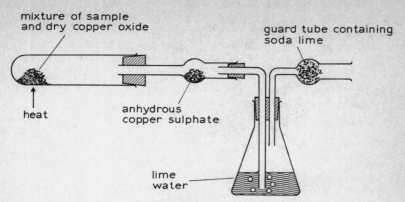

mixture of sample and dry copper oxide

guard tube containing soda lime

heat

anhydrous copper sulphate

lime water

Fig. 1 Detection of carbon and hydrogen

reaction ceases. The red-hot tube is then plunged into an evaporating dish half filled with distilled water, care being taken to provide adequate shielding from possible explosion. After the reaction with water, the contents of the evaporating dish are crushed with a glass rod, and then heated just to boiling as an aid to dissolving the substances formed during fusion. The insoluble material is filtered off and the filtrate is used for subsequent tests. The products formed as a result of the sodium fusion are shown in Table 1.

TABLE 1 *The inorganic products formed from an organic compound during sodium fusion*

| Element present in organic compound | Fused with | Product |
|---|---|---|
| C \\ N | Na | NaCN |
| S | Na | $Na_2S$ |
| Cl | Na | NaCl |
| Br | | NaBr |
| I | | NaI |
| P | Na (and atmospheric oxygen) | $Na_3PO_4$ |

The presence of these products is confirmed as follows:

**Cyanide**   A small quantity of iron (II) sulphate solution is added to a portion of the filtrate. A thick greenish-grey precipitate of iron (II) hydroxide should form. (If this does not happen the fusion procedure

11

must be repeated using a slightly larger piece of sodium, or a smaller quantity of organic material.) The mixture is then boiled, during which process some of the iron (II) ions are oxidized to iron (III). The mixture is acidified with dilute sulphuric acid, and a deep blue colour caused by a fine suspension of Prussian blue precipitate shows the presence of a cyanide. From this it is concluded that nitrogen is present in the original organic compound.

$$Fe^{2+} + 6CN^- = [Fe(CN)_6]^{4-}$$
$$Fe^{3+} + Na^+ + [Fe(CN)_6]^{4-} = NaFe[Fe(CN)_6]$$

**Sulphide** A few drops of sodium pentacyano-nitroso-ferrate (III) solution are added to a portion of the filtrate. A violet colour, which disappears on warming, confirms the presence of a sulphide, which in turn shows the presence of sulphur in the original organic material.

**Halide** A third portion of the filtrate is made acid to litmus with dilute nitric acid. If either sulphide or cyanide is present, a little more nitric acid is added and the solution is boiled to expel all the hydrogen sulphide or hydrogen cyanide. Formation of a precipitate on the addition of silver nitrate solution indicates the presence of a halide. From the colour of this precipitate and its behaviour on the addition of aqueous ammonia it is possible to say which halide is present.

The presence of a chloride is shown by a white precipitate which dissolves in aqueous ammonia and which reappears on the addition of dilute nitric acid:

$$Ag^+ + Cl^- = AgCl$$
$$AgCl + 2NH_3 \xrightarrow[HNO_3]{NH_3} [Ag(NH_3)_2]^+ + Cl^-$$

Bromide is indicated by a slightly off-white precipitate which is not readily soluble in aqueous ammonia solution:

$$Ag^+ + Br^- = AgBr$$

Iodide gives a cream precipitate which is insoluble in aqueous ammonia solution:

$$Ag^+ + I^- = AgI$$

Both sulphide and cyanide ions interfere with this test as the corresponding silver salts are insoluble. Silver cyanide is also white and it is very easily confused with silver chloride, so that it is important to remove these interfering ions by the preliminary boiling with dilute nitric acid.

**Phosphate** A small portion of the filtrate is made acid by the addition of concentrated nitric acid. The solution resulting from this acidification should be quite warm; otherwise it is heated to about 60°C. An excess of ammonium molybdate solution is added, and the presence of phosphate is shown by the formation of a yellow *crystalline* precipitate. A positive result in this test shows the presence of phosphorus in the organic compound.

**Metals** A small quantity of the original organic compound is placed on a crucible lid which is then strongly heated. Heating is continued until all the carbon has been burnt away. The formation of a residue (often white) of metal oxide indicates the presence of a metal in the organic compound. The metal may be identified by the usual inorganic qualitative techniques.

### Quantitative Analysis

Knowing which elements are present in a compound, the next step is to determine their relative proportions using the methods of quantitative analysis.

A high degree of accuracy in the quantitative analysis is called for, in view of the fact that the relative proportions of the elements may differ only slightly between one compound and another. This is achieved by using carefully purified reagents and samples of the compound under test, and the analysis is usually carried out on a semi-micro scale using between 25 and 50 mg of the starting material. Briefly, the methods used are as follows:

**Carbon and hydrogen** The estimation of these elements is carried out in an apparatus similar to that shown in Figure 2, using a combustion method. Purified oxygen is admitted to the combustion chamber and

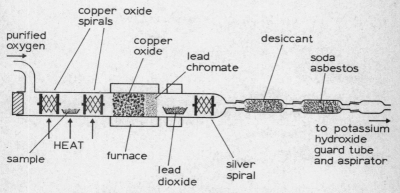

Fig. 2 Quantitative combustion analysis apparatus

passes over two heated copper oxide spirals. A porcelain boat containing a known weight of the sample is placed between the two spirals, and the portion of the combustion chamber within the electrically heated furnace is filled with copper oxide. At the end of this section is a packing of lead chromate which reacts with, and so retains, oxides of sulphur. Finally, a charge of lead dioxide, which is kept at a lower temperature than the rest of the furnace by an independent heater, is included in order to absorb any oxides of nitrogen which are produced during the oxidation of organic compounds containing nitrogen. When halogens are present, a silver spiral is introduced after the lead dioxide boat. The gases produced by combustion are passed through two weighed absorption tubes, the first of which contains a desiccant, often magnesium perchlorate, and the second soda–asbestos (sodium hydroxide held on asbestos gauze). The increases in the weights of these two tubes give the weights of water and carbon dioxide formed during combustion.

From these results, the proportions of carbon and hydrogen present in the sample can be calculated as shown in the following example:

EXAMPLE   0·0245 g of an organic compound produced 0·04563 g of carbon dioxide and 0·02827 g of water on combustion analysis. Taking the relative atomic masses of carbon, hydrogen and oxygen as 12·01, 1·008 and 16·00 respectively, calculate the percentage of carbon and hydrogen present in the compound.

44·01 g of carbon dioxide contain 12·01 g of carbon. Thus 0·04563 g of carbon dioxide contain

$$\frac{12 \cdot 01 \times 0 \cdot 04563}{44 \cdot 01} = 0 \cdot 01245 \text{ g carbon}$$

The proportion of carbon in the compound is

$$\frac{0 \cdot 01245 \times 100}{0 \cdot 0245} = 50 \cdot 8\%$$

Similarly, 18·02 g of water contain 2·016 g of hydrogen. Thus 0·02827 g of water contain

$$\frac{2 \cdot 016 \times 0 \cdot 02827}{18 \cdot 02} = 0 \cdot 003163 \text{ g hydrogen}$$

The proportion of hydrogen in the compound is

$$\frac{0 \cdot 003163 \times 100}{0 \cdot 0245} = 12 \cdot 9\%.$$

**Halogens and sulphur**  These elements are estimated by the *Carius* method. A weighed quantity of the organic compound is placed in a stout glass tube together with fuming nitric acid and a few silver nitrate crystals. The tube is then sealed, wrapped in asbestos paper (to prevent the glass being scratched) and heated in a strongly constructed iron furnace to 200°C for six hours. After cooling, the tube is opened cautiously and the silver halide precipitate is filtered, washed, dried and weighed. The following example illustrates the calculation involved.

EXAMPLE  0·1446 g of an organic liquid known to contain chlorine produced 0·5203 g of pure dry silver chloride during a Carius determination.

The weight of chlorine in the sample of organic compound is

$$\frac{35\cdot46 \times 0\cdot5203}{143\cdot4} = 0\cdot1287 \text{ g}$$

Thus the proportion of chlorine in the compound is

$$\frac{0\cdot1287 \times 100}{0\cdot1446} = 89\cdot0\%$$

When this method is used for the estimation of sulphur in an organic compound, the silver nitrate is replaced by barium nitrate, and the final precipitate formed under these conditions is barium sulphate. The calculation is similar to that given for the chlorine determination, the weight of sulphur present in the given sample of the organic compound being equal to $32\cdot06 \times w/233\cdot5$ where $w$ is the weight of the barium sulphate precipitate.

**Nitrogen: the Dumas method**  This is one of two methods available for the quantitative determination of nitrogen in an organic compound. It may be used for all nitrogen-containing organic compounds. A known weight of the compound is well mixed with copper (II) oxide and the mixture is strongly heated in the combustion apparatus shown in Figure 3. The combined nitrogen is converted into gaseous nitrogen (or oxides of nitrogen which are later reduced to elementary nitrogen by the hot copper gauze located further along the combustion tube), while carbon dioxide and water are also produced by the combustion of the organic compound. A slow stream of carbon dioxide is fed into the apparatus, and the nitrogen is collected in a nitrometer over concentrated potassium hydroxide solution (this absorbs the carbon dioxide and condenses the water vapour). The volume of nitrogen collected in the nitrometer (corrected to s.t.p. and

15

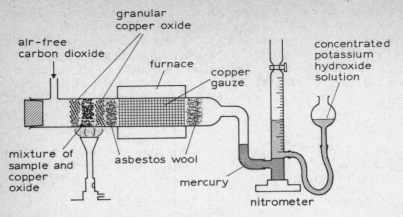

Fig. 3 Dumas apparatus for nitrogen determination

for the presence of water vapour) indicates the quantity of nitrogen combined in the organic material.

EXAMPLE    0·0280 g of an organic compound produced 5·60 cm³ of nitrogen, measured dry and corrected to s.t.p. in a Dumas determination. The density of nitrogen is 0·00125 g cm⁻³ at s.t.p. 5·60 cm³ nitrogen weigh 5·60 × 0·00125 g. Thus the proportion of nitrogen in the compound is

$$\frac{5\cdot60 \times 0\cdot00125 \times 100}{0\cdot0280} = 25\cdot0\%$$

**Nitrogen: the Kjeldahl method**    This is quicker and simpler than the Dumas method, but it does not give satisfactory results for all nitrogen-containing organic compounds (pyridine for example). The method is based on the fact that nitrogen in an organic compound can often be converted into ammonia by the prolonged action of hot concentrated sulphuric acid, the end product being ammonium sulphate. This part of the process is carried out in a long-necked Kjeldahl flask (Figure 4a), and after this treatment, the acid solution is poured into a large round-bottomed distillation flask containing a quantity of water. The apparatus is set up as shown in Figure 4b and the ammonia is liberated by the addition of concentrated sodium hydroxide solution. The liberated ammonia is distilled over into a known volume of standard acid (the splash head serves to prevent any sodium hydroxide spray from being carried over into the standard acid). The amount of ammonia liberated is determined by back titrating the excess acid with standard alkali.

16

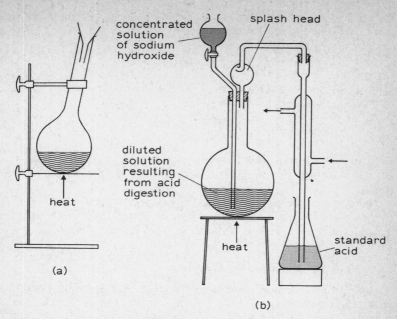

Fig. 4 Kjeldahl nitrogen determination

(a) Long-necked Kjeldahl flask in which the organic compound is digested with concentrated sulphuric acid

(b) Ammonia distillation apparatus used in the second stage of the Kjeldahl determination

EXAMPLE   The ammonia produced from 0·200 g of a compound in a Kjeldahl determination was absorbed in 20·0 cm³ of M hydrochloric acid. At the end of the determination, the excess acid required 18·5 cm³ M sodium hydroxide solution for neutralization. From these results, it is clear that the ammonia produced neutralized

$$20·0 - 18·5 = 1·50 \text{ cm}^3 \text{ M hydrochloric acid}$$

Now 1000 cm³ M hydrochloric acid by definition contains 1 mole hydrogen chloride. Therefore, 1·50 cm³ M hydrochloric acid contains

$$\frac{1 \times 1·50}{1000} \text{ moles HCl}$$

Since $NH_3 + HCl = NH_4Cl$, 1 mole hydrochloric acid is neutralized by 1 mole ammonia, so that 1·50/1000 moles of ammonia must have

17

been produced. But 1 mole ammonia contains 14·01 g nitrogen; thus the organic compound contains by weight

$$\frac{1 \cdot 50 \times 14 \cdot 01}{1000} = 0 \cdot 021 \text{ g nitrogen}$$

Hence the proportion of nitrogen in the compound is

$$\frac{0 \cdot 021 \times 100}{0 \cdot 200} = 10 \cdot 5\%$$

**Oxygen**  The proportion of oxygen present in an organic compound is usually estimated by difference after accounting for all the other elements.

There are many variations and improvements on these classical techniques. Modern methods incorporate instrumental techniques, among which the mass spectrometer is especially important.

### Calculating the Empirical Formula

An empirical formula represents the simplest whole-number ratio in which the atoms are present in a molecule of the compound. The proportions by weight of each element present in a certain compound have been determined by quantitative analysis. Each of these proportions is now divided by the relative atomic mass of the element concerned and the result shows the ratio of the numbers of atoms of each element present. This ratio is seldom a whole-number ratio; to obtain this, each term in the ratio is divided by the smallest. The following example illustrates a typical calculation.

EXAMPLE  Combustion analysis of 0·176 g of an organic compound known to contain carbon, hydrogen, and possibly oxygen, produced 0·198 g of water and 0·242 g of carbon dioxide.

The weight of carbon present in 0·242 g of carbon dioxide is

$$\frac{12 \cdot 01 \times 0 \cdot 242}{44 \cdot 01} = 0 \cdot 06604 \text{ g}$$

Expressed as a percentage, the compound contains

$$\frac{0 \cdot 06604 \times 100}{0 \cdot 176} = 37 \cdot 5\% \text{ carbon}$$

The weight of hydrogen present in 0·198 g water is

$$\frac{2 \cdot 016 \times 0 \cdot 198}{18 \cdot 01} = 0 \cdot 02216 \text{ g}$$

18

and, as a percentage,

$$\frac{0 \cdot 02216 \times 100}{0 \cdot 176} = 12 \cdot 6\%$$

By difference, the proportion of oxygen present in the compound is

$$100 - (37 \cdot 5 + 12 \cdot 6) = 49 \cdot 9\%$$

| Element | Percentage | $\dfrac{Percentage}{Relative\ atomic\ mass}$ | Whole number ratio |
|---|---|---|---|
| C | 37·5 | $\dfrac{37 \cdot 5}{12 \cdot 01} = 3 \cdot 120$ | $\dfrac{3 \cdot 120}{3 \cdot 119} = 1 \cdot 00$ |
| H | 12·6 | $\dfrac{12 \cdot 6}{1 \cdot 008} = 12 \cdot 5$ | $\dfrac{12 \cdot 5}{3 \cdot 119} = 4 \cdot 01$ |
| O | 49·9 | $\dfrac{49 \cdot 9}{16 \cdot 00} = 3 \cdot 119$ | $\dfrac{3 \cdot 119}{3 \cdot 119} = 1 \cdot 00$ |

Slight deviations ($\pm 1\%$) in the whole numbers in the final column may be ignored since these are due to slight errors in the quantitative analysis. Thus, the empirical formula for this compound is $CH_4O$.

When the figures in the final column are significantly different from whole numbers, such deviations cannot be attributed to errors in experimental technique, and further steps must be taken to obtain a whole-number ratio, as shown in the following example.

EXAMPLE A hydrocarbon is found to contain $81 \cdot 80\%$ carbon and $18 \cdot 20\%$ hydrogen. Dividing these percentages by the respective relative atomic masses, we get

$$C = \frac{81 \cdot 80}{12 \cdot 01} = 6 \cdot 811$$

$$H = \frac{18 \cdot 20}{1 \cdot 008} = 18 \cdot 05$$

On dividing through by the smallest number, a whole-number ratio does not appear:

$$C = \frac{6 \cdot 811}{6 \cdot 811} = 1 \cdot 00$$

$$H = \frac{18 \cdot 05}{6 \cdot 811} = 2 \cdot 65$$

The deviation from a whole-number ratio is too large to be accounted for by experimental error, and the empirical formula is obtained by

19

taking the lowest multiple of the above ratio which produces a whole-number ratio; i.e.,

$$C:H = 1:2{\cdot}65 \quad \text{or} \quad 1:2\tfrac{2}{3}$$

which when multiplied by 3 gives 3:8.

The empirical formula for this hydrocarbon is $C_3H_8$.

## Determination of Molecular Formulae

The empirical formula expresses the simplest ratio of atoms which agrees with the percentage composition of the compound, but it does not indicate the actual number of atoms present in the molecule. The *molecular* formula gives this information, and this formula is either identical with, or is a multiple of, the empirical formula.

EXAMPLE  The empirical formula $CH_2O$ corresponds to

| | | |
|---|---|---|
| methanal | $CH_2O$ | (HCHO) |
| ethanoic acid | $C_2H_4O_2$ | ($CH_3CO_2H$) |
| methyl methanoate | $C_2H_4O_2$ | ($HCOOCH_3$) |
| glucose | $C_6H_{12}O_6$ | etc. |

The appropriate molecular formula is obtained from a knowledge of the relative molecular mass of the compound. The relative molecular mass can be determined from measurements of vapour density, osmotic pressure, elevation of boiling point, depression of freezing point and from mass spectrometry. Further details of most of these techniques can be found in Part 3.

Taking the case of a compound with empirical formula $CH_2O$, and assuming that it is found to have a relative molecular mass of 60, then

$$(CH_2O)_n = n(12 + 2 + 16) = 60$$

from which

$$n = 2$$

(A high degree of accuracy is not called for here, as it is only necessary to distinguish whether $n$ is one, two or three, etc.) The molecular formula for this compound is, therefore, $(CH_2O)_2$ or $C_2H_4O_2$.

As we have seen, this molecular formula corresponds to two different substances, ethanoic acid and methyl methanoate. The molecules of each of these isomers have different structures, that is the atoms combined in the molecule are linked in a different sequence in each case. Consequently, the next stage in the investigation of an organic compound is the deduction of its *structural* formula.

## Elucidation of Structural Formulae

A great deal of evidence has to be collected, using both physical and chemical methods of investigation, before it is possible to suggest a possible structural formula for a compound. Some of the physical methods include:

(a) *X-ray diffraction methods.* Both X-rays and light are electromagnetic radiations, although the wavelength of X-rays is very much shorter than that of light rays. When a beam of X-rays of a single wavelength is passed through a crystal, a diffraction pattern is produced which depends on the relative positions of the atoms within the crystal. With the help of a computer, the patterns resulting from complicated substances can be interpreted. From the results, an electron density map can be constructed which allows the positions of the several atoms in the compound to be deduced. In most cases, it is not possible to locate hydrogen atoms by this technique, although similar methods using electron or neutron beams are more successful; however, such experiments are more complex.

(b) *Nuclear magnetic resonance.* This method is based on the fact that the hydrogen atoms absorb small quantities of radiant energy when a substance containing hydrogen is placed in a very intense magnetic field. The exact wavelengths of the radiation absorbed depend on the relative positions of the hydrogen atoms in the molecule, and thus give some information regarding the numbers and positions of the hydrogen atoms in the molecule. Certain other elements, fluorine and phosphorus for example, also exhibit this phenomenon.

(c) *Mass spectrometry.* In addition to the determination of isotopic masses (see Part 3), this technique can be helpful in investigating molecular structure. When electrons of successively higher energies are used to bombard the molecule, the compound breaks down into successively smaller fragments. The identity of these fragments is detected by the instrument and from this information it is often possible to piece together the structural detail of the molecule.

(d) *Infrared spectrophotometry.* This method is based on the fact that a molecule can absorb energy in the infrared region of the electromagnetic spectrum resulting in an increase in its vibrational energy. In simple terms, vibration can be visualized as the shortening and the extension of a bond, and the energy associated with this process depends on the strength of the bond and the relative masses of the atoms joined by it. Thus, the different types of bond linking various atoms absorb infrared radiation at different characteristic frequencies and the presence of these bonded atoms can be deduced from an infrared absorption spectrum.

(e) *Polarimetry*. This is the measurement of the extent of the rotation of the plane of polarized light by optically active isomers (pp. 137–9).

(f) *Chromatography*. This is a technique which has proved invaluable in the detection, separation and identification of compounds. Briefly, chromatography depends on the different degrees of retention that obtain between various substances and the chromatographic medium (called the stationary phase). In paper chromatography, a special type of absorbent paper is used. Even in normal, dry paper, a certain amount of moisture is retained within the fibres, and it is these moist fibres which form the stationary phase.

In Figure 5 a mixture of two substances, A and B, dissolved in a suitable solvent is placed in the form of a spot at the top of a sheet of chromatography paper. After the evaporation of the solvent, another solvent is allowed to flow down the paper, and the mixture separates to give the spots A and B as shown.

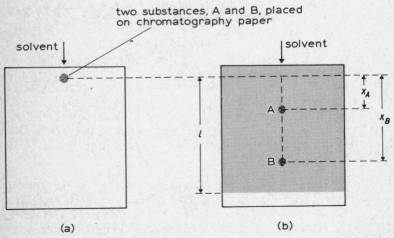

Fig. 5 Formation of a chromatogram

Compounds held more strongly by the stationary phase travel more slowly in the flowing solvent than a substance which is retained less firmly. This is expressed in terms of an $R_f$ value, which is defined by the equation

$$R_f = \frac{\text{distance moved by the sample}}{\text{distance moved by the solvent front}}$$

In Figure 5(b) the solvent has moved a distance $l$, and the $R_f$ values for A and B respectively are

$$\frac{x_A}{l} \quad \text{and} \quad \frac{x_B}{l}$$

Alternatively, the moving solvent may be allowed to climb the sheet of paper (ascending chromatography) or flow outwards from the centre of a paper circle (circular chromatography). In all these cases, the moving solvent carries with it the constituents of the mixture at different rates and a satisfactory separation of the components of the mixture is usually achieved. When colourless substances are present, the location of the spots on the final chromatogram is made possible by spraying the paper with a chemical which produces a colour with the various spots (this is called *developing* the chromatogram) or by viewing the chromatogram in ultraviolet light. In practice, several control spots of known substances are run side by side with the unknown mixture. If, under the conditions of the experiment, the chromatogram shows that one of the control spots has the same $R_f$ value as one of the spots from the mixture, then the identity of one component is fairly certain. The value of paper chromatography as a method of dealing with small samples of material at ordinary temperatures and capable of giving reliable results was further enhanced by the introduction of *two-way chromatography*. In this technique, a normal chromatogram is first obtained using either an ascending or descending solvent. It is quite possible that substances of a similar chemical nature will fail to separate under these conditions. To overcome this, the chromatogram is dried and a second solvent is made to flow over it in a direction at right angles to that in which the first solvent was run. As a result of this treatment with two solvents, a much more complete separation is obtained. *Thin layer chromatography* is rather similar to paper chromatography. A paste of alumina, silica gel etc. is applied to give a thin coating to a glass plate. When dried, this thin layer of material forms the stationary phase (replacing the paper in paper chromatography) and the technique is very little different from that already described. Thin layer chromatography generally gives a sharper separation of spots, and the range of stationary phases that can be employed extends the usefulness of the method.

In *column chromatography*, the stationary phase is packed in a column (up to 3 or 4 cm in diameter) and a small quantity of the solution to be analysed is placed at the top of the packing. The components of the mixture are carried through the packing by a solvent (known as an eluant) and separate into distinct bands as they descend the column. By continuing the solvent flow, each band can be washed through the column completely and so each component of the mixture can be collected separately.

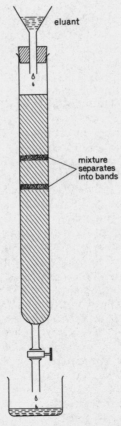

eluant

mixture separates into bands

Fig. 6 Column chromatography

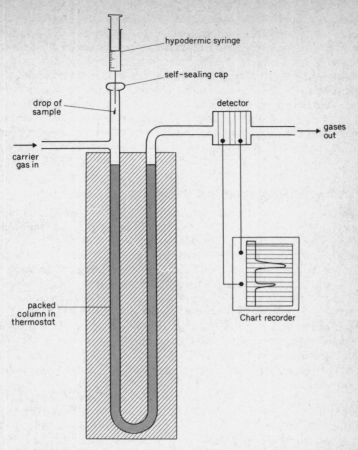

Fig. 7 Simplified flow diagram of a gas–liquid chromatography unit

Column chromatography can also be applied to the analysis of gases and volatile liquid mixtures. In this type of chromatography, known as *gas–liquid* or *vapour phase chromatography*, the stationary phase is packed into a narrow metal tube (often of considerable length) which is housed in a cabinet and maintained at a carefully controlled temperature. The liquid solvent used in column chromatography is replaced by a carrier gas such as hydrogen or nitrogen, and the sample is injected into the system by means of a hypodermic syringe pushed through a self-sealing rubber cap as shown in Figure 7. A detector is fitted in the gas stream at the end of the column, and the purpose of this device is to produce an electrical signal, which can be fed to a

25

chart recorder so that a peak is introduced in the moving trace whenever a substance other than the carrier gas passes through the detector. The time between the injection of the sample and the appearance of a peak on the trace is called the retention time, and every compound, under a given set of conditions has a characteristic retention time. Usually, conditions are chosen so that the retention times range from fifteen or twenty seconds to two or three minutes. This technique, which is capable of analysing rapidly small samples of complex mixtures is extensively used in industry (particularly in oil refineries and petrochemical plants), in medicine and in research.

### Homologous Series, Functional Groups and Organic Radicals

Many of these physical techniques are of recent introduction, while the original elucidation of the structure of many organic compounds was based on a careful study of the chemical reactions of the compound. Examples of the steps leading to the assignment of a structural formula from, in the main, chemical evidence, are given in later sections, for some of the simpler organic compounds.

Various combinations of atoms—called *functional groups*—when present in a molecule, give rise to characteristic chemical behaviour. Much of the chemistry included in the following chapters describes the general reactions of several functional groups, and it is only when we are familiar with the types of behaviour associated with each functional group that it is possible to suggest a structural formula for a compound.

A vast number of organic compounds are known. Fortunately, their study is made easier by considering the chemistry of the series, or families, into which they may be classified. The classification is made according to the functional group, which, as indicated above, is the group or part of a molecule which determines the principal reactions of the molecule. Such a series of compounds is termed a *homologous series*, and each series shows the following characteristics:

1 All the compounds in the series have a common functional group.
2 All the compounds in the series show similar chemical behaviour, and their reactions may be summarized by *general reactions*.
3 The chemical constitution of all the members of the series can be expressed by means of a *general formula*.
4 The relative molecular masses of successive members of the series show a regular increment, while the physical properties of the successive members show a regular gradation in value.

Members of the same homologous series are known as *homologues*.

The simplest organic compounds—the *alkanes*—are saturated hydrocarbons, and they are regarded as not having a functional group; rather they are assumed to act as *parent* compounds, in that they provide a *radical* (root) to which the functional group may be attached. An example will make this clearer. Ethane, a parent hydrocarbon, has the formula $C_2H_6$, while chlorine can act as a functional group. Replacing one of the hydrogen atoms in ethane by chlorine leads to the formation of chloro-ethane, $C_2H_5Cl$, in which the $C_2H_5-$ part is the radical, and the $-Cl$ is the functional group.

The general formula for the alkanes is $C_nH_{2n+2}$, and each parent hydrocarbon may have a hydrogen atom replaced by a functional group X, to produce $C_nH_{2n+1}X$. The radical $C_nH_{2n+1}$ is known as an *alkyl* (from alkane) radical, and is sometimes given the symbol R for general representation.

# 3 The Aliphatic Hydrocarbons

Hydrocarbons are compounds containing hydrogen and carbon only. When each carbon atom in the chain is linked to four other atoms (i.e. four single bonds) the hydrocarbon is said to be *saturated*. In *unsaturated* hydrocarbons, neighbouring carbon atoms in the chain are linked by double or triple bonds.

## THE ALKANES

Alkanes form the major constituents of crude petroleum which is at present one of the world's major raw materials. In addition to its use as a fuel, oil is the basis of a very important section of the chemical industry (Chapter 18). Alkanes are also present in natural gas from underground wells; North Sea gas, for example, contains a high proportion of methane.

The general formula for this homologous series is $C_nH_{2n+2}$

TABLE 2 *The first members of the alkane series*

| Name and formula | M.P. °C | B.P. °C | Relative molecular mass | Radical |
|---|---|---|---|---|
| methane $CH_4$ | −184 | −163 | 16 | methyl $CH_3-$ |
| ethane $CH_3CH_3$ | −172 | − 87 | 30 | ethyl $CH_3CH_2-$ |
| propane $CH_3CH_2CH_3$ | −188 | − 43 | 44 | propyl $CH_3CH_2CH_2-$ |
| butane $CH_3CH_2CH_2CH_3$ | −135 | − 0·5 | 58 | butyl $CH_3CH_2CH_2CH_2-$ |

## General Methods of Preparation

1 Heat the sodium salt of the corresponding acid with soda lime:

$$R.CO_2^-Na^+ + NaOH \rightarrow R.H + Na_2CO_3$$

The $CO_2^-$ group is called the *carboxyl* group, and during this reaction it is destroyed. Hence this type of reaction is termed *decarboxylation*. Soda lime is made by treating calcium oxide with sodium hydroxide solution, and it is preferred to sodium hydroxide as it is not deliquescent.

2 The reaction of a Grignard reagent (p. 47) with water:

$$R.MgX + H_2O \rightarrow R.H + Mg(OH)X$$

The reaction is quite vigorous even in the cold.

3 By the reduction of an alkyl halide using zinc and dilute acid:

$$R.X + Zn + H^+ \rightarrow R.H + Zn^{2+} + X^-$$

4 Reduction of an aldehyde or ketone using amalgamated zinc and concentrated hydrochloric acid:

$$R.CHO + 2Zn + 4HCl \rightarrow R.CH_3 + 2Zn^{2+} + H_2O + 4Cl^-$$

$$R.CO.R' + 2Zn + 4HCl \rightarrow R.CH_2.R' + 2Zn^{2+} + H_2O + 4Cl^-$$

This reaction is known as a Clemmensen reduction.

## Special Methods of Preparation

1 *The Wurtz reaction.* An iodo-alkane (dissolved in dry ethoxy-ethane (ether)) is refluxed over sodium. The resulting alkane is collected from the top of the reflux condenser if the alkane is gaseous, or by distillation after the reaction in the case where the alkane is a liquid:

$$2R.I + 2Na \rightarrow 2NaI + R.R$$

Good yields may be obtained by this method when alkanes having an even number of carbon atoms are produced, but a mixture of two different iodo-alkanes gives rise to three alkanes in the product. The Wurtz reaction produces an alkane having a greater number of carbon atoms in the chain than the starting material, so that this reaction affords a method of *ascending a homologous series.* (Note that with decarboxylation, one carbon atom is lost from the starting material, and this reaction is one whereby a *descent* of the series may be made.)

2 *Special method for methane.* The action of water on aluminium carbide. When water or dilute acid is added drop by drop to aluminium carbide, methane is generated.

$$Al_4C_3 + 12H_2O \rightarrow 3CH_4 + 4Al(OH)_3$$

## Properties and Reactions of the Alkanes

The lower members are colourless, odourless gases, while the higher members are liquids or waxes, often possessing pleasant odours. They are insoluble in water, but dissolve in most organic solvents.

The alkanes are characterized by their general unreactivity. The following reactions of methane are typical of the series.

1 It burns readily in air:

$$CH_4 + 2O_2 \rightarrow CO_2 + 2H_2O$$

In a limited supply of air, the presence of unburnt carbon in the product causes a yellow, sooty flame:

$$CH_4 + O_2 \rightarrow 2H_2O + C$$

2 When mixed with chlorine, and on exposure to diffuse light, a substitution product is formed:

$$CH_4 + Cl_2 \rightarrow HCl + CH_3Cl \qquad \text{chloromethane}$$

together with smaller quantities of products ($CH_2Cl_2$, $CHCl_3$ and $CCl_4$) formed by the successive replacement of hydrogen atoms by chlorine. In these reactions, a hydrogen atom is replaced (or substituted) by a chlorine atom; hence the reactions are termed *substitution* reactions. This type of reaction is typical of saturated hydrocarbons. Because a mixture of all four of the above products results from the reaction of chlorine and methane under these conditions, this reaction is an unsatisfactory method of preparing these substances on a laboratory scale.

## Introduction to the Theory of Organic Reactions

A chemical reaction between covalent molecules involves the rupture and the re-making of bonds. There are two ways in which a bond (i.e. a shared electron pair) may break:

1 Each atom may retain one electron of the shared pair

$$A:B \rightarrow A\cdot + \cdot B$$

The products are *free radicals* and the process is termed *homolytic fission* of the bond.

2 The more electronegative atom may retain both of the electrons which form the shared pair.

$$A:B \rightarrow A^+ + :B^-$$

(or vice-versa if A is more electronegative than B). The products are *ions* and the process is termed *heterolytic fission*.

A careful study of the reactions of a certain substance and a knowledge of the conditions under which the reaction takes place can provide evidence to show whether homolytic or heterolytic fission has taken place. Briefly, when a reaction is speeded up by the absorption of energy (such as radiant energy, especially ultraviolet rays) or by the presence of substances which are known to produce free radicals easily, we presume that homolytic fission occurs. On the other hand, when a reaction takes place more readily in the presence of substances which help ion formation (such as polar solvents, acids or bases) heterolytic fission is thought to be involved.

## Substitution of the Alkanes

The chlorination of methane is a photochemical reaction and the stages are as follows:

1 Initiation step. Chlorine atoms are produced from chlorine molecules on the absorption of energy from ultraviolet light:

$$Cl:Cl + energy \rightarrow Cl\cdot + \cdot Cl$$

2 Propagation steps.

$$CH_4 + Cl\cdot \rightarrow HCl + CH_3\cdot$$

$$Cl_2 + CH_3\cdot \rightarrow CH_3Cl + Cl\cdot$$

The sequence continues, thus setting up a self-maintained or *chain reaction*. This theory accounts for the spread of products formed during this reaction, since the free chlorine atoms may attack chloromethane (or dichloromethane, etc.) molecules in addition to methane in the second of the above stages.

The term *reaction mechanism* is given to a series of stages which trace the course of a reaction.

## Isomerism

This is the phenomenon of two or more compounds having the same molecular formula. Butane is the first alkane which exhibits isomerism, since it is possible to arrange the chain of carbon atoms in two ways:

(a)
```
    H   H   H   H
    |   |   |   |
H—C—C—C—C—H
    |   |   |   |
    H   H   H   H
```

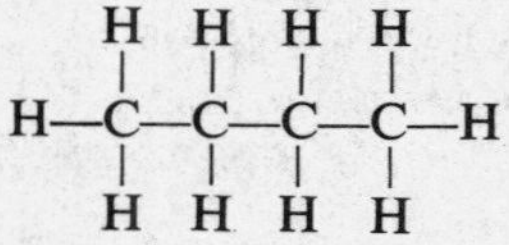

(b)
```
    H   H   H
    |   |   |
H—C—C—C—H
    |   |   |
    H   |   H
        |
      H—C—H
        |
        H
```

31

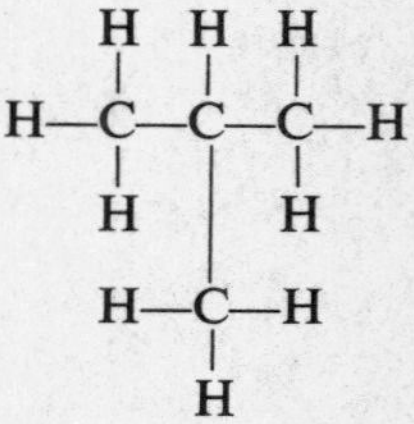

(a) A straight chain of four carbon atoms. This isomer is called butane, boiling point $-0.5°C$.

(b) A branched structure in which the longest chain contains three carbon atoms. This isomer is called methyl propane and it has a boiling point of $-12°C$.

The number of isomers increases rapidly with an increase in the number of carbon atoms in the molecule; for example, butane has two isomers, pentane three and octane eighteen.

## Names of the Alkanes

The rules for naming organic compounds are those set out by the International Union of Pure and Applied Chemistry (the IUPAC system). In the case of the alkanes, the name of each member ends in *-ane*. The part of the name preceding this ending depends on the number of carbon atoms in the longest unbranched carbon chain in the molecule. Table 3 indicates the application of these rules.

TABLE 3

| No. of carbon atoms in chain | Designated by | Name of alkane | Name of alkyl radical |
|---|---|---|---|
| 1 | meth- | methane | methyl |
| 2 | eth- | ethane | ethyl |
| 3 | prop- | propane | propyl |
| 4 | but- | butane | butyl |
| 5 | pent- | pentane | pentyl |
| 6 | hex- | hexane | hexyl |
| 7 | hept- | heptane | heptyl |
| 8 | oct- | octane | octyl |
| 9 | non- | nonane | nonyl |
| 10 | dec- | decane | decyl |

The *systematic* name (i.e. that given according to the IUPAC rules) is found by writing the structure of the alkane so that the longest continuous carbon chain is evident, and any branches or side chains are named as alkyl radicals. For example, the systematic name for iso-butane (iso-butane being what is known as a trivial name) is methyl propane, since the longest carbon chain contains three carbon atoms, and the methyl (alkyl) group is attached to the second carbon atom in the propane chain:

$$CH_3—CH—CH_3$$
$$|$$
$$CH_3$$

When two or more different alkyl groups are attached to the main chain, these are named in alphabetical order (i.e. ethyl before methyl, etc.) and the numbering of the carbon atoms in the main chain is such that the number of the carbon atom to which the side chain is attached is kept as low as possible. The following examples illustrate these rules.

(a) The isomer of hexane having the structural formula as shown is named 2,2-dimethylbutane (not 3,3-dimethylbutane); the name is found by numbering from right to left.

$$CH_3-CH_2-\underset{\underset{\displaystyle CH_3}{|}}{\overset{\overset{\displaystyle CH_3}{|}}{C}}-CH_3$$

(b) The isomer of nonane with the following structural formula is 4-ethyl-2-methylhexane and not 2-methyl-4-ethylhexane, since the alkyl groups are listed in alphabetical order.

$$CH_3-\underset{\underset{\displaystyle CH_3}{|}}{CH}-CH_2-\underset{\underset{\displaystyle CH_2}{|}\atop\underset{\displaystyle CH_3}{|}}{CH}-CH_2-CH_3$$

## The Formula of Gaseous Hydrocarbons by Eudiometry

**Outline of the experiment** A measured volume of the hydrocarbon is mixed with a large excess of oxygen, and the total volume is noted. The mixture is then sparked until combustion is complete. After cooling to the original conditions of temperature and pressure, and when all the water formed by combustion has condensed, the volume of the remaining gas is measured. The volume of carbon dioxide produced is found from the diminution in the total volume after shaking with concentrated potassium hydroxide solution. As a test for complete combustion, the oxygen may be absorbed in alkaline pyrogallol, after which treatment no gas should remain.

If the gaseous hydrocarbon has the general formula $C_xH_y$, the equation representing complete combustion of this hydrocarbon will be:

$$C_xH_y + \left(x + \frac{y}{4}\right)O_2 \rightarrow xCO_2 + \frac{y}{2}H_2O$$

If the volumes of the products and reactants are measured under the same conditions of temperature and pressure, then the volumes of the gases are proportional to the number of molecules of gas shown in the equation. That is to say, if the initial volume of the hydrocarbon used is $V$ cm$^3$, we can multiply all the quantities in the equation by $V$ to obtain the volumes of each substance taking part in the reaction:

$$V.C_xH_y + V\left(x + \frac{y}{4}\right)O_2 \rightarrow Vx.CO_2 + V\left(\frac{y}{2}\right)H_2O$$

Thus, $V$ cm$^3$ of hydrocarbon produce $Vx$ cm$^3$ of carbon dioxide and $V(y/2)$ cm$^3$ of steam, which on cooling to the initial temperature condenses to water, the volume of which is ignored.

From the results of an experiment, the value of $x$ is found from the volume of carbon dioxide produced, while the value of $y$ is deduced from the overall loss in volume on combustion. An example shows the method of calculation.

10 cm$^3$ of a gaseous hydrocarbon were mixed with 100 cm$^3$ of oxygen and the mixture was sparked. On cooling to the original conditions of temperature and pressure, the volume measured 85 cm$^3$, of which 20 cm$^3$ was absorbed by concentrated potassium hydroxide solution.

*Calculation.* In this case, $V = 10$ cm$^3$. The volume of carbon dioxide produced is

$$20 \text{ cm}^3 = Vx \text{ cm}^3$$

from the above reasoning. Thus,

$$10\,x = 20$$
$$x = 2$$

The value of $y$ is determined from the loss in volume on combustion.

$$\text{Total initial volume} = V + Vx + V\frac{y}{4} \text{ cm}^3$$

$$\text{Final volume} = Vx \text{ cm}^3$$

Any excess volume of oxygen cancels when calculating the diminution in volume. Loss in volume on combustion is

$$V + Vx + \frac{Vy}{4} - Vx = V + \frac{Vy}{4} \text{ cm}^3$$

In this example, the loss in volume is

$$10 + \frac{10y}{4} = 110 - 85 = 25 \text{ cm}^3$$

Thus

$$\frac{10y}{4} = 25 - 10 = 15$$

$$10y = 60$$

$$y = 6$$

The formula of the hydrocarbon is thus $C_2H_6$

## THE ALKENES

The general formula for the alkenes is $C_nH_{2n}$ and the functional group may be regarded as the double bond which links two neighbouring carbon atoms in alkenes. The alkenes are, therefore, *unsaturated* hydrocarbons.

TABLE 4 *The lower members of the alkene series*

| Name | Formula | B.P. °C |
|------|---------|---------|
| ethene (ethylene) | $CH_2{=}CH_2$ | $-105$ |
| propene (propylene) | $CH_2{=}CHCH_3$ | $-\ 48$ |
| but-1-ene | $CH_2{=}CHCH_2CH_3$ | $-\ \ 6$ |
| but-2-ene | $CH_3CH{=}CHCH_3$ | $1$ |
| methylpropene | $CH_3\overset{\displaystyle |}{\underset{\displaystyle CH_3}{C}}{=}CH_2$ | $-\ \ 7$ |

The lower members of the alkene series are not found to any large extent in nature. Some alkenes of high relative molecular mass are found in plants.

### General Methods of Preparation

1 Alkenes may be produced by the dehydration of alcohols. Dehydration may be achieved by heating with concentrated sulphuric acid at 165°C (the yield is not good owing to side product formation), or with syrupy (i.e. concentrated) phosphoric acid at 225°C. A good yield of alkene is obtained when alcohol vapour is passed over alumina at 350°C.

$$R.CH_2CH_2OH \rightarrow H_2O + R.CH{=}CH_2$$

2 On boiling an alkyl halide with a concentrated, alcoholic solution of potassium hydroxide an alkene is produced.

$$R.CH_2CH_2Br + KOH \rightarrow KBr + H_2O + R.CH{=}CH_2$$

The yield varies with the type of alkyl halide used.

35

**Manufacture** Ethene, propene and butene are widely used as the starting point for the production of other organic chemicals, plastics and rubber, etc., and they are manufactured in large quantities from petroleum sources (pp. 161–4).

## Properties and Reactions of the Alkenes

The lower members of this series are colourless gases, and they are insoluble in water.

In contrast to the alkanes, the alkenes are highly reactive substances. The reactivity is due to the presence of the double bond, and the characteristic reaction is *addition*, which is the combination of two compounds to form a single product.

1 Ethene burns in air:

$$C_2H_4 + 3O_2 \rightarrow 2CO_2 + 2H_2O$$

2 Hydrogen, in the presence of a nickel catalyst adds on to alkenes forming the corresponding alkane. The reaction takes place at 200°C. For example:

$$CH_2{=}CHCH_3 + H_2 \rightarrow CH_3CH_2CH_3$$

3 Chlorine and bromine add on to alkenes readily, but reaction with iodine does not take place. For example, at room temperature ethene discharges the colour of bromine and produces an oily liquid, 1,2-dibromoethane:

$$CH_2{=}CH_2 + Br_2 \rightarrow CH_2BrCH_2Br$$

4 Concentrated aqueous solutions of hydrogen bromide and hydrogen iodide add on readily to double bonds:

$$R.CH{=}CH_2 + HX \rightarrow R.\underset{\underset{X}{|}}{C}HCH_3$$

The products formed in this reaction are predicted by *Markovnikov's rule*, which states:

*The addition of a hydrogen halide to an alkene, under normal conditions, gives a product formed by the addition of hydrogen to the carbon atom, which already carries the greater number of hydrogen atoms.*

5 Hypochlorous acid adds on to alkenes to give chlorhydrins:

$$R.CH{=}CH_2 + HOCl \rightarrow R.CHOHCH_2Cl$$

a chlorhydrin

36

6 Cold dilute solutions of potassium permanganate (in either acid or alkaline conditions) are rapidly decolourized by alkenes. The initial product is a *diol* (a compound containing two hydroxy-groups) which may in turn be attacked by the permanganate.

$$R.CH{=}CH_2 + H_2O + (O) \rightarrow R.CHOHCH_2OH$$

7 Alkenes dissolve in cold, concentrated sulphuric acid to form alkyl hydrogen sulphates:

$$R.CH{=}CH_2 + H_2SO_4 \rightarrow R.CH(SO_4H)CH_3$$

alkyl hydrogen sulphate

Since the alkanes do not react with cold, concentrated acid, this is a useful method of removing alkenes from alkanes.

8 Ozone reacts with alkenes to form unstable compounds known as ozonides:

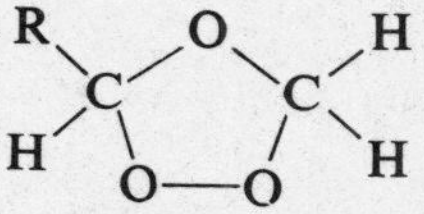

On hydrolysis of an ozonide two molecules of carbonyl compounds (either aldehydes or ketones) are produced. The identification of these hydrolysis products indicates the position of the double bond in the alkene. This useful method of analysis is termed *ozonolysis*

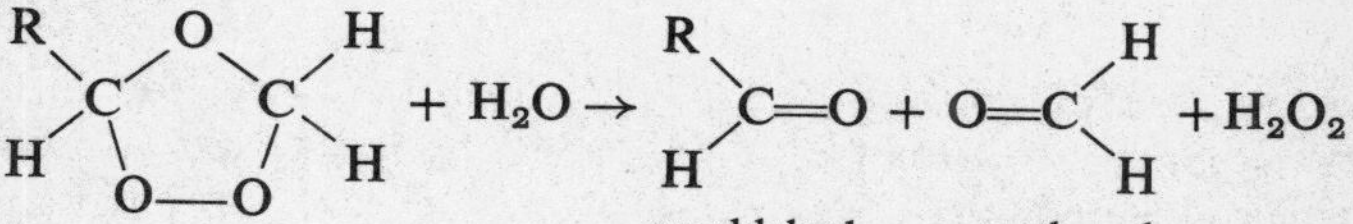

an aldehyde    methanal

9 Under conditions of high pressures and temperatures, the double bonds in alkene molecules may break, and a succession of alkene molecules may link together. This reaction is termed *polymerization*, the product is a *polymer*, and the starting material is termed the *monomer*. With ethene, the polymer resulting is polythene:

$$nCH_2{=}CH_2 \rightarrow (-CH_2-CH_2-)_n$$

The polymer molecules have a very high relative molecular mass, although their empirical formula is the same as that of the monomer.

An alternative process, the Ziegler process, is used for the production of polythene, in which the use of high temperatures and pressures is eliminated by the employment of a catalyst (often aluminium and titanium compounds). The polythene produced by this method is more rigid and has a higher density than that produced by the high pressure method, but it is usually discoloured since it contains traces of the catalyst retained within the product.

## Theory of the Addition Reactions of Alkenes

In the alkenes, two electron pairs are shared between two adjacent carbon atoms, thus forming a double bond. Experiment shows that one of the two bonds is more readily broken than the other, and this leads to the special reactivity of the alkenes. Modern electronic theory of valency (see Part 1 of this series) also indicates that the bonds are of unequal strength. To distinguish the bonds, one of them is termed a sigma ($\sigma$) bond—this is the stronger one—and the other is known as a pi ($\pi$) bond. The reactions of the alkenes are centred on the rupture of the weaker $\pi$ bond. This bond undergoes heterolytic fission in the presence of an attacking molecule since:

(a) alkenes undergo addition reactions in the absence of light;
(b) addition reactions are catalysed by the presence of polar substances; for example, dry ethene and dry chlorine in a wax or polythene vessel do not react until a trace of water is added;
(c) when bromine adds on to ethene in a solution containing chloride ions, the expected product, $CH_2BrCH_2Br$, also contains $CH_2BrCH_2Cl$.

In the case of the addition of hydrogen bromide, the bromine atom is more electronegative than the hydrogen atom, so that hydrogen bromide is a polar molecule. On the approach of the hydrogen bromide molecule both electrons in the $\pi$ bond move towards one carbon atom.

The addition of the hydrogen atom to the partially negatively charged carbon atom is the first step in the addition, as the two $\pi$-electrons can be donated to the proton formed by the heterolytic fission of the H—Br bond. The second state is the addition of the bromide ion to the positive carbon as shown in Figure 8.

Similar mechanisms can be traced for the addition of sulphuric acid and bromine (which is capable of polarization to $Br^{\delta+}$—$Br^{\delta-}$).

It appears that the first link to be formed between the alkene and the attacking molecule is the one involving the negative carbon atom.

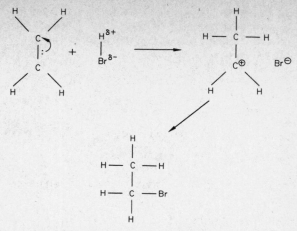

Fig. 8 Addition of hydrogen bromide to ethene

Consequently, the attacking molecule must, in some way, promote the formation of a negative carbon and then be able to bond with the pair of electrons which becomes available. Molecules that are capable of effecting this are known as *electrophilic* or electron-seeking reagents. This theory also accounts for the results of addition predicted by Markovnikov's rule. In ethene, a symmetrical molecule, either carbon atom can accept the $\pi$-electron pair. In unsymmetrical molecules, propene, for example, the movement of the electron pair is governed by the relative electron-repelling effects of the atoms or groups attached to the two carbon atoms linked by the double bond. Now alkyl groups have a greater electron-repelling power than hydrogen atoms, so that the $\pi$-electron pair moves away from the alkyl group and the final products are in agreement with the Markovnikov rule.

$$CH_3 . CH{=}CH_2 + HX \rightarrow CH_3 . CHX . CH_3$$

## Names of the Alkenes

The systematic name given to the members of the alkene series is derived by using the name-ending -ene coupled with the forename given according to the number of carbon atoms present, as shown in Table 3 on p. 32. In addition to this, the position of the double bond is shown by interposing the number of the first carbon atom joined by the double bond between the forename and the name ending. For example, $CH_3CH_2CH{=}CH_2$ is but-1-ene and not but-3-ene, since

39

the IUPAC convention requires that the numbers be kept as small as possible by reversing the order of numbering if necessary.

Branched chain alkenes are named in a similar way to the alkanes. As an example,

$$CH_3-\overset{\overset{\displaystyle CH_3}{|}}{CH}-CH=CH-CH_2-CH_3$$

is 2-methylhex-3-ene.

**Isomerism** arises among the alkenes in three distinct ways:

(a) through the different arrangement of groups, as with the alkanes,

(b) through the different positions of the double bond in the alkene.

Thus the alkane containing four carbon atoms has two isomeric forms (butane and methyl propane), while the corresponding alkene has three isomeric forms, as shown in Table 4 on p. 35.

(c) geometrical isomerism (see p. 140).

## THE ALKYNES

The alkynes have the general formula $C_nH_{2n-2}$ and they all contain a triple bond which may be regarded as the functional group in this class of hydrocarbons. The names, structural formula and boiling points of the early members of the series are given in Table 5.

TABLE 5 *The lower members of the alkyne series*

| Name | Structural formula | B.P. °C |
|---|---|---|
| ethyne (or acetylene) | $HC\equiv CH$ | $-83\cdot4$ |
| propyne | $CH_3C\equiv CH$ | $-23\cdot3$ |
| but-1-yne | $HC\equiv CCH_2CH_3$ | $8\cdot5$ |
| but-2-yne | $CH_3C\equiv CCH_3$ | $27\cdot1$ |
| pent-1-yne | $HC\equiv CCH_2CH_2CH_3$ | $39\cdot6$ |
| pent-2-yne | $CH_3C\equiv CCH_2CH_3$ | $55\cdot3$ |
| 3-methylbut-1-yne | $HC\equiv CC\overset{\overset{\displaystyle |}{H}}{\underset{\underset{\displaystyle CH_3}{|}}{}}CH_3$ | $27\cdot8$ |

### Names of the Alkynes

The IUPAC system of nomenclature is similar to that described for the other aliphatic hydrocarbons except that the names of the members of this series end in -*yne*.

As with the alkenes, isomerism arises on account of the different arrangements of the groups in the main carbon chain, and the different positions taken up by the triple bond. The triple bond does not give rise to geometrical isomerism, since the four atoms in the system X—C≡C—X lie in a straight line.

## General Method of Preparation

By the reaction of concentrated potassium hydroxide solution on a *vic*-dihalide (i.e. a halogen derivative of a hydrocarbon which has two halogen atoms linked to adjacent carbon atoms in the chain) in boiling alcohol:

$$R—CH—CH—R' + 2KOH \rightarrow 2KBr + 2H_2O + R—C≡C—R'$$
$$\underset{Br}{|} \quad \underset{Br}{|}$$

where R and R' represent the same, or different alkyl groups.
   *Vic*-dihalides are usually prepared from an alkene:

$$R—CH=CH—R' + Br_2 \rightarrow R—CH—CH—R'$$
$$\underset{Br}{|} \quad \underset{Br}{|}$$

so that this is a method for converting alkenes into alkynes.

## Special Method of Preparation

Ethyne (acetylene) is prepared by the action of water on calcium carbide:

$$CaC_2 + 2H_2O \rightarrow Ca(OH)_2 + HC≡CH$$

**Manufacture**   Ethyne is manufactured from natural gas, and used in large quantities for the production of polymers.

## Properties and Reactions of the Alkynes

These hydrocarbons are colourless, insoluble in water, and have a boiling point some 15° to 20°C above the corresponding alkenes. Ethyne is usually associated with an objectionable odour, although it is reported to have a faint ethereal smell when pure. The gas is highly explosive when compressed, so that it is transported as a solution in acetone under moderate pressure.
   Being highly unsaturated, the chief reactions of the alkynes are addition and polymerization, chemical attack being centred on the

41

triple bond. The most important alkyne is ethyne, its reactions will be described as they are typical of the series.

1 Ethyne burns readily in air, often with a very sooty flame:

$$2C_2H_2 + 5O_2 \rightarrow 4CO_2 + 2H_2O$$

A mixture of oxygen and ethyne explodes violently when ignited. When ethyne is burnt in oxygen, a very hot flame (up to 3000°C) is produced.

2 Ethyne can be made to combine with hydrogen in two stages.

(a) In the presence of the correct quantity of hydrogen and using a special palladium catalyst, ethene is formed:

$$HC\equiv CH + H_2 \rightarrow CH_2{=}CH_2$$

(b) In the presence of a heated nickel or platinum catalyst, and with excess hydrogen, ethane is produced:

$$HC\equiv CH + 2H_2 \rightarrow CH_3CH_3$$

3 Ethyne explodes when mixed with chlorine, forming carbon and hydrogen chloride. In the presence of an inert material as a diluent, tetrachloroethane is produced:

$$HC\equiv CH + 2Cl_2 \rightarrow CHCl_2CHCl_2$$

Tetrachloroethane is a poisonous, non-inflammable liquid. It is readily converted into trichloroethene (trichloroethylene), which has a lower toxicity and is widely used as a solvent in the dry-cleaning industry.

Bromine and iodine react progressively less vigorously with ethyne.

4 Hydrogen iodide and hydrogen bromide combine with ethyne in two stages:

(a) $$HC\equiv CH + HI \rightarrow CH_2{=}CHI$$
$$\text{vinyl iodide}$$

(b) $$CH_2{=}CHI + HI \rightarrow CH_3CHI_2$$

The second addition takes place according to Markovnikov's rule.

Hydrogen chloride reacts much less readily, but in the presence of a catalyst, the gases react at 175°C:

$$HC\equiv CH + HCl \rightarrow CH_2{=}CHCl$$

The product, vinyl chloride, is unsaturated and undergoes polymerization to form polyvinyl chloride (PVC), which is a well-known plastic:

$$nCH_2=CHCl \rightarrow (—CH_2CH—)_n$$
$$\phantom{nCH_2=CHCl \rightarrow (—CH_2CH—)_n} |$$
$$\phantom{nCH_2=CHCl \rightarrow (—CH_2CH—)} Cl$$

Further addition of hydrogen chloride produces 1,1-dichloro-ethane:

$$CH_2=CHCl + HCl \rightarrow CH_3CHCl_2$$

5 When bubbled into warm dilute sulphuric acid containing mercury (II) sulphate catalyst, ethanal is formed:

$$HC \equiv CH + H_2O \rightarrow CH_3CHO$$

6 The terminal hydrogen atom in alkynes of the type $R.C \equiv CH$ is acidic and can be replaced by a metal. Ethyne, for example, gives an off-white precipitate of silver dicarbide when bubbled through an ammoniacal solution of $Ag^+$

$$2Ag^+ + HC \equiv CH \rightarrow 2H^+ + Ag—C \equiv C—Ag$$

Copper (I) dicarbide is similarly formed as a red precipitate when ethyne is passed through an ammoniacal solution of copper (I) chloride.

These covalent compounds, unlike the ionic calcium carbide, are explosive when dry.

7 Polymerization. Ethyne polymerizes in several different ways, depending on the conditions under which the reaction takes place.

When passed through a heated metal tube, benzene is produced in low yield:

$$3C_2H_2 \rightarrow C_6H_6$$

# 4 Halogen Compounds

## MONO-HALOGENO-ALKANES (ALKYL HALIDES)

The general formula for halogeno-alkanes is $C_nH_{2n+1}X$, where $X$ represents a halogen atom. As the name of the series implies, the compounds are derived from the substitution of a hydrogen atom in an alkane by a halogen. The chemical reactivity of these compounds is centred on the halogen atom, which is the functional group.

TABLE 6 *Lower mono-halogeno-alkanes*

| Name | Formula | M.P. °C | B.P. °C |
|------|---------|---------|---------|
| chloromethane | $CH_3Cl$ | $-97$ | $-23\cdot8$ |
| bromomethane | $CH_3Br$ | $-93$ | $4\cdot6$ |
| iodomethane | $CH_3I$ | $-64$ | $42\cdot4$ |
| chloroethane | $CH_3CH_2Cl$ | $-138$ | $12\cdot8$ |
| bromoethane | $CH_3CH_2Br$ | $-118$ | $38\cdot3$ |
| iodoethane | $CH_3CH_2I$ | $-111$ | $72\cdot4$ |
| 1-chloropropane | $CH_3CH_2CH_2Cl$ | $-123$ | $46\cdot7$ |
| 2-chloropropane | $CH_3\overset{|}{C}H.CH_3$ $\underset{Cl}{}$ | $-117$ | $36\cdot7$ |

### General Methods of Preparation

Direct substitution of a hydrogen atom in an alkane by a halogen is not a good preparative method. Alcohols are the most convenient starting points.

1 Bromo- and iodo-alkanes may be prepared in moderately good yields by refluxing the appropriate alcohol with concentrated hydrobromic or hydriodic acid. The yields of bromide are increased in the presence of concentrated sulphuric acid.

$$R.OH + HX \rightarrow R.X + H_2O$$

Chloro-alkanes may be produced by passing gaseous hydrogen chloride into the alcohol in the presence of anhydrous zinc chloride.

2 Chloro-alkanes may be made by treating an alcohol with phosphorus pentachloride, generally at room temperature.

$$R.OH + PCl_5 \rightarrow R.Cl + POCl_3 + HCl$$

For the bromo- and iodo-alkanes the technique is different; freshly dried red phosphorus is suspended in the alcohol and either iodine or bromine is added in small quantities at a time, after which the mixture is refluxed in order to produce a reasonable yield of alkyl halide.

$$3R.OH + PX_3 \rightarrow 3R.X + H_3PO_3$$

3 From alkenes by the addition of a hydrogen halide (p. 36):

$$R.CH{=}CH_2 + HX \rightarrow R.\underset{\underset{X}{|}}{C}HCH_3$$

**Manufacture**  Chloromethane is manufactured either by the reaction between hydrogen chloride and methanol, or by the direct chlorination of methane at an elevated temperature, using a large excess of methane.

Chloroethane is produced by the direct chlorination of ethane at temperatures over 350°C, with the ethane present in large excess:

$$CH_3CH_3 + Cl_2 \rightarrow CH_3CH_2Cl + HCl$$

Linked with this process is a second which utilizes the hydrogen chloride produced in this reaction to form a further quantity of chloroethane by reaction with ethene:

$$CH_2{=}CH_2 + HCl \rightarrow CH_3CH_2Cl$$

### Properties and Reactions of the Halogeno-Alkanes

These compounds are gases or oily liquids, which have a characteristic sweet smell. They are practically insoluble in both water and cold concentrated sulphuric acid, thus allowing them to be separated from alkenes and alcohols.

The reactivity of the halogeno-alkanes depends on the reactions of the functional group—the halogen atom. In general, the reactivity of the halogen group increases in the order: chloride, bromide, iodide, corresponding to the gradual decrease in the carbon–halogen bond strength.

$$C{-}F \quad C{-}Cl \quad C{-}Br \quad C{-}I$$

The bond between carbon and fluorine is extremely strong, consequently organic fluoro-alkanes are unreactive.

### Reaction Mechanism of the Halogeno-Alkanes

In a halogeno-alkane molecule, the electron pair forming the carbon–halogen bond will be pulled by the halogen atom away from the carbon. This gives the carbon atom a slight positive charge and any species which is to form a bond with the carbon must be attracted by the positive charge. Such species are known as nucleophiles; typical examples being negative ions such as $OH^-$, $CN^-$ and molecules such as ammonia which possess a lone pair of electrons. The mechanism of the attack of a halogeno-alkane, $R.CH_2X$ by $OH^-$ can be represented as follows:

$$HO^- : \underset{\underset{R}{|}}{\overset{\overset{H}{|}}{C}} - X \rightarrow HO - \underset{\underset{H}{|}}{\overset{\overset{H}{|}}{C}} - R + X^-$$

The curved arrow on the left shows that a lone pair from the hydroxide ion moves to form a bond with the carbon, and at the *same instant* the pair of electrons forming the C—X bond transfer to the halogen to form the departing $X^-$ ion. This reaction is a substitution reaction in which the attacking species is a nucleophile. A convenient notation is to designate this as an $S_N2$ reaction (substitution by a nucleophile) which requires the presence of two molecules to determine the reaction (i.e. it is second order—see Part 3). Cyanide ion ($:CN^-$) displaces the halogen in a similar manner; the reaction is carried out in alcoholic solution.

A complicating factor is that some halogeno-alkanes react by means of a different mechanism. In some cases, the carbon–halogen bond undergoes heterolytic fission as a slow first stage in the reaction. The second stage is the rapid attachment of a nucleophile to the positive carbon atom so formed. The rate of this reaction is governed by the speed at which the carbon–halogen bond ruptures, and as this process involves one molecule only, the substitution is designated $S_N1$.

$$-\overset{|}{\underset{|}{C}} - X \rightarrow -\overset{|}{\underset{|}{C}}{}^+ + :X^- \quad \text{slow first stage}$$

$$-\overset{|}{\underset{|}{C}}{}^+ + :Z^- \rightarrow -\overset{|}{\underset{|}{C}} - Z \quad \text{rapid attachment of nucleophile } Z^-$$

## Summary of Reactions

The reactions of bromoethane are given as being typical of the series.

1 It is reduced to the parent alkane by sodium amalgam:

$$CH_3CH_2Br + H_2O + 2(Na) \rightarrow CH_3CH_3 + NaBr + NaOH$$

2 On reaction with dry sodium in ethoxy-ethane, it is converted to a higher alkane (this is the *Wurtz reaction*):

$$2CH_3CH_2Br + 2Na \rightarrow CH_3CH_2CH_2CH_3 + 2NaBr$$

3 On boiling with dilute potassium hydroxide (or a suspension of silver oxide in water), the parent alcohol is formed:

$$CH_3CH_2Br + KOH \rightarrow KBr + CH_3CH_2OH$$

4 An alkene is produced on boiling under reflux with a concentrated alcoholic solution of potassium hydroxide. With bromoethane, the yield of ethene is very small, although higher yields of alkene are obtained from higher alkyl halides:

$$CH_3CH_2CH_2Br + KOH \rightarrow CH_3CH{=}CH_2 + KBr + H_2O$$

5 Ethers are formed by the reaction between alkyl halides and sodium alkoxides (this reaction is *Williamson's synthesis*):

$$CH_3CH_2Br + R.ONa \rightarrow NaBr + CH_3CH_2O.R$$

6 Refluxing using a solution of potassium cyanide in alcohol produces a nitrile:

$$CH_3CH_2Br + KCN \rightarrow CH_3CH_2CN + KBr$$

A similar reaction using silver cyanide results in the formation of an isonitrile:

$$CH_3CH_2Br + AgCN \rightarrow CH_3CH_2NC + AgBr$$

7 An amine salt is produced by the reaction between a halide and a concentrated alcoholic solution of ammonia:

$$CH_3CH_2Br + NH_3 \rightarrow CH_3CH_2NH_3{}^+Br^-$$

ethylammonium bromide

Further attack can take place (p. 94)

8 Refluxing a halogeno-alkane in dry ethoxy-ethane with dry magnesium granules gives an alkyl magnesium halide—a *Grignard reagent*:

$$CH_3CH_2Br + Mg \rightarrow CH_3CH_2MgBr$$

Grignard reagents are highly reactive compounds, and they are readily converted into alkanes, acids and alcohols (p. 166).

Grignard compounds are classified as *organo-metallic* compounds, in which the metal and carbon atoms are linked by a covalent bond. Lead (IV) tetraethyl (used in making 'anti-knock' additives for petrol) is another organo-metallic compound.

It is manufactured by heating chloroethane with a lead–sodium alloy in a pressure vessel:

$$4CH_3CH_2Cl + 4Na/Pb \rightarrow 3Pb + 4NaCl + (CH_3CH_2)_4Pb$$

## OTHER HALOGEN COMPOUNDS

### Trichloromethane [Chloroform] $CHCl_3$

#### Preparation

Trichloromethane is one of the products resulting from the substitution of hydrogen in methane by chlorine.

The above reaction is not suitable for the preparation of trichloromethane owing to the mixture of products obtained, and the difficulty of their separation. It is best prepared by the *haloform reaction*, which depends on the reaction of chlorine with an alkaline solution of ethanol, ethanal or propanone. Bleaching powder is used to provide both the chlorine and the alkali needed for the reaction. The reaction is thought to take place in the following stages:

1 *From ethanol*

(a) Oxidation of ethanol to ethanal:

$$CH_3CH_2OH + Cl_2 \rightarrow 2HCl + CH_3CHO$$

(b) Chlorination of ethanal:

$$CH_3CHO + 3Cl_2 \rightarrow 3HCl + CCl_3CHO$$

(c) Reaction of the product with alkali:

$$CCl_3CHO + OH^- \rightarrow HCO_2^- + CHCl_3$$

2 *From ethanal*; the reaction sequence begins at stage (b) above.

3 *From propanone*

(a) Chlorination of propanone

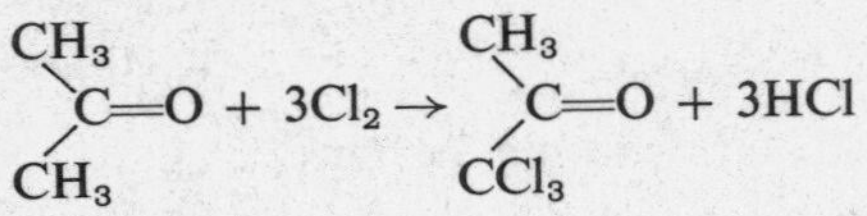

$$\begin{array}{c} CH_3 \\ \diagdown \\ \phantom{xx}C{=}O + 3Cl_2 \rightarrow \\ \diagup \\ CH_3 \end{array} \begin{array}{c} CH_3 \\ \diagdown \\ \phantom{xx}C{=}O + 3HCl \\ \diagup \\ CCl_3 \end{array}$$

(b) Reaction with alkali:

$$\begin{array}{c} CH_3 \\ \diagdown \\ \phantom{xx}C{=}O + OH^- \rightarrow CH_3CO_2^- + CHCl_3 \\ \diagup \\ CCl_3 \end{array}$$

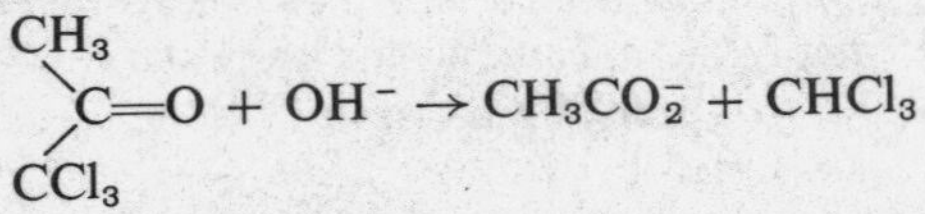

# Properties and Reactions of Trichloromethane

Trichloromethane is a colourless, sweet-smelling liquid, which is almost insoluble in water. It is an anaesthetic but now it is mainly used as a solvent and as a starting point for the production of some fluorinated hydrocarbons.

It has the following chemical properties:

1 The liquid is not inflammable, but the vapour burns with a green-edged flame.

2 It is slowly decomposed by air in bright sunlight:

$$2CHCl_3 + O_2 \rightarrow 2HCl + 2COCl_2$$

carbonyl chloride

Carbonyl chloride is a highly toxic gas.

3 It is decomposed by refluxing with concentrated sodium hydroxide solution:

$$CHCl_3 + 4NaOH \rightarrow 3NaCl + 2H_2O + HCO_2Na$$

4 When heated with phenylamine and potassium (or sodium) hydroxide solution, an evil-smelling product, phenyl isonitrile, is formed:

$$CHCl_3 + 3KOH + C_6H_5NH_2 \rightarrow 3KCl + 3H_2O + C_6H_5NC$$

phenyl isonitrile

This reaction provides a very sensitive test for trichloromethane, and it is an example of a general reaction that takes place between trichloromethane and any primary amine:

$$CHCl_3 + 3KOH + R.NH_2 \rightarrow 3KCl + 3H_2O + R.NC$$

## Tribromomethane [Bromoform] CHBr$_3$

This is a colourless liquid which has a very high specific gravity (2·8), and it is used in flotation methods of mineral separation. It is made by the haloform reaction using bromine and an alkaline solution of ethanol, ethanal or propanone.

## Tri-iodomethane [Iodoform] CHI$_3$

Tri-iodomethane is a pale yellow solid, having a pleasant 'antiseptic' odour. It is produced by the reaction between iodine and an alkaline solution of ethanol, ethanal or propanone. The stages are similar to those shown on p. 48. The overall reactions are:

49

$$CH_3CH_2OH + 4I_2 + OH^- \rightarrow 5HI + HCO_2^- + CHI_3$$

$$CH_3CHO + 3I_2 + OH^- \rightarrow 3HI + HCO_2^- + CHI_3$$

$$\begin{array}{c} CH_3 \\ \diagdown \\ C=O + 3I_2 + OH^- \rightarrow 3HI + CH_3CO_2^- + CHI_3 \\ \diagup \\ CH_3 \end{array}$$

The formation of tri-iodomethane is used to distinguish between ethanol and methanol, since methanol does not give this reaction, and to test for ethanal, as this is the only aldehyde which can react in this way. Propanone, or any methyl ketone, can form tri-iodomethane. (*Note:* the reaction is commonly referred to as the *iodoform reaction* or *test.*)

Tri-iodomethane is decomposed by alkali in a similar way to trichloromethane. The solid melts at 120°C and decomposes to give off iodine on heating to higher temperatures. It is also decomposed slightly by warm silver nitrate solution, and a faint precipitate of silver iodide is produced.

## Tetrachloromethane [Carbon Tetrachloride] $CCl_4$

This is a colourless, sweetish smelling non-inflammable liquid. It is widely used as a solvent, in fire extinguishers and in the production of fluorinated hydrocarbons.

It is manufactured by reacting carbon disulphide with chlorine in the presence of an anhydrous iron (III) chloride catalyst:

$$CS_2 + 3Cl_2 \rightarrow CCl_4 + S_2Cl_2$$

Tetrachloromethane is not very reactive. It is oxidized to carbonyl chloride especially when in contact with air at a hot metal surface—a factor which limits the use of this substance as a fire extinguisher. On passing anhydrous hydrogen fluoride into tetrachloromethane, using antimony pentachloride as a catalyst, dichlorodifluoromethane is formed.

$$CCl_4 + 2HF \xrightarrow{SbCl_5} CCl_2F_2 + 2HCl$$

## Fluoro-Compounds

Fluoro-alkanes are prepared by indirect methods, since direct fluorination of the alkanes usually leads to the decomposition of the molecule. Satisfactory methods are:

1 The addition of hydrogen fluoride to an alkene; for example:

$$CH_2=CH_2 + HF \rightarrow CH_3CH_2F$$

2 The replacement of chlorine in a chlorinated alkane by fluorine using hydrogen fluoride or a suitable inorganic fluoride; for example:

$$CHCl_3 + 2HF \rightarrow 2HCl + CHClF_2$$
chlorodifluoromethane

$$2CH_3CH_2Br + HgF_2 \rightarrow 2CH_3CH_2F + HgBr_2$$

$$CCl_4 + 2HF \rightarrow 2HCl + CCl_2F_2$$

Compounds having more than one fluorine atom linked to the same carbon are particularly stable and inert. Dichlorodifluoromethane is an ideal refrigerant; it is non-toxic, non-corrosive, non-inflammable, almost odourless, boils at $-30°C$ and does not decompose below 500°C. It has been employed as a refrigerant since 1930, and it is now widely used as the propellant gas in aerosol sprays.

Tetrafluoroethene is made by heating chlorodifluoromethane to 700°C.

$$2CHClF_2 \rightarrow CF_2{=}CF_2 + 2HCl$$

This unsaturated compound polymerizes on heating under pressure in the presence of ammonium persulphate, to produce polytetrafluoroethene (PTFE). PTFE is a plastic which is chemically inert, heat-resistant (softening above 300°C) and which has the lowest coefficient of friction of any known substance. It is widely used in non-lubricated bearings and for non-stick coatings.

# 5   The Monohydric Alcohols

The general formula for the monohydric alcohols is $C_nH_{2n+1}OH$ (or R.OH). They may be regarded as hydroxyl derivatives of the alkanes. The term *monohydric* indicates the presence of a single hydroxyl group (—OH) in the molecule.

TABLE 7 *Lower monohydric alcohols*

| Name | Formula | M.P.°C | B.P.°C |
|------|---------|--------|--------|
| methanol (methyl alcohol) | $CH_3OH$ | − 97 | 64·4 |
| ethanol (ethyl alcohol) | $CH_3CH_2OH$ | −114 | 78·4 |
| propan-1-ol (n-propyl alcohol) | $CH_3CH_2CH_2OH$ | −125 | 97·3 |
| propan-2-ol (iso-propyl alcohol) | $CH_3CHCH_3$<br>$\quad\ \ OH$ | − 88 | 82·3 |
| butan-1-ol (n-butyl alcohol) | $CH_3CH_2CH_2CH_2OH$ | − 90 | 117·5 |
| butan-2-ol (sec-butyl alcohol) | $CH_3CHCH_2CH_3$<br>$\quad\ \ OH$ | | 99·8 |
| 2-methylpropan-1-ol (iso-butyl alcohol) | $CH_3$<br>$\quad\ \ CHCH_2OH$<br>$CH_3$ | −108 | 108 |
| 2-methylpropan-2-ol (tert-butyl alcohol) | $CH_3\quad CH_3$<br>$\quad C$<br>$CH_3\quad OH$ | 25 | 82·6 |

The functional group of the alcohols is the hydroxyl group.

## Types of Alcohol: Primary, Secondary and Tertiary

Alcohols are classified according to the location of the hydroxyl group in the molecule.

*Primary alcohols* are those in which the hydroxyl group forms part of a —CH₂OH group. That is, the hydroxyl group is linked to a carbon atom, which is itself joined to not more than one other carbon atom. Methanol, H—$CH_2OH$, and 2-methylpropan-1-ol,

$$CH_3CHCH_2OH$$
$$|$$
$$CH_3$$

are both examples of primary alcohols.

*Secondary alcohols* are distinguished by the presence of a $\diagdown CHOH$ group in which the hydroxyl group is linked to a carbon atom which is itself joined to two other carbon atoms. Propan-2-ol,

$$CH_3$$
$$\diagdown$$
$$\diagup CHOH$$
$$CH_3$$

is an example of a secondary alcohol.

*Tertiary alcohols* contain the group $\overset{|}{\underset{|}{-C-}}OH$, in which the carbon atom carrying the hydroxyl group is also linked to three other carbon atoms as in 2-methylpropan-2-ol,

$$CH_3$$
$$|$$
$$CH_3-C-CH_3$$
$$|$$
$$OH$$

These different classes of alcohol arising as a result of differences in the position of the hydroxyl group in the molecule are examples of a type of isomerism known as *position isomerism*. The products and the rate at which the reactions of these classes of alcohols take place may differ significantly.

## General Methods of Preparation

1 Hydrolysis of a halogeno-alkane. Water slowly decomposes many alkyl halides, but the use of hot dilute alkali gives a better yield of alcohol:

$$R.X + KOH \rightarrow KX + R.OH$$

2 Hydration of alkenes. The alkene is allowed to react with sulphuric acid (p. 37) and the product is hydrolysed:

$$R.CH{=}CH_2 \xrightarrow{H_2SO_4} R.CHCH_3 \xrightarrow{H_2O} R.CHCH_3$$
$$\qquad\qquad\qquad\quad |\qquad\qquad\qquad |$$
$$\qquad\qquad\qquad\quad SO_4H\qquad\qquad\quad OH$$

3 Reduction of an aldehyde, ketone or carboxylic acid using sodium amalgam and water (or by catalytic hydrogenation). Aldehydes yield primary alcohols:

$$R.CHO \xrightarrow{2H} R.CH_2OH$$

Ketones produce secondary alcohols:

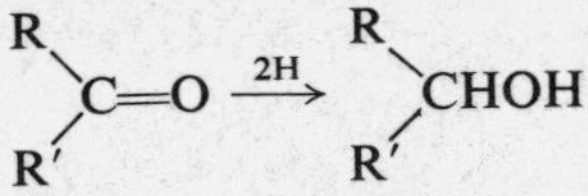

$$\begin{matrix} R \\ \diagdown \\ & C{=}O \xrightarrow{2H} \\ \diagup \\ R' \end{matrix} \quad \begin{matrix} R \\ \diagdown \\ & CHOH \\ \diagup \\ R' \end{matrix}$$

Acids give rise to primary alcohols in low yield using lithium aluminium hydride:

$$R.CO_2H \xrightarrow{4H} R.CH_2OH$$

Tertiary alcohols cannot be made by reduction.

4 From aldehydes and ketones, using a Grignard reagent. For example, a Grignard reagent reacts with propanone as follows:

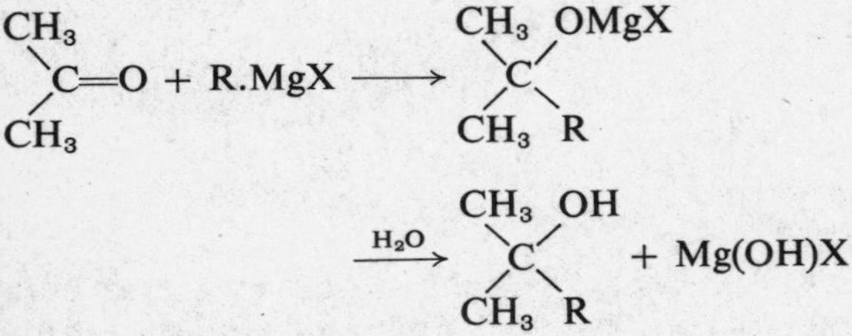

$$\begin{matrix} CH_3 \\ \diagdown \\ & C{=}O + R.MgX \longrightarrow \\ \diagup \\ CH_3 \end{matrix} \quad \begin{matrix} CH_3 \quad OMgX \\ \diagdown \diagup \\ C \\ \diagup \diagdown \\ CH_3 \quad R \end{matrix}$$

$$\xrightarrow{H_2O} \quad \begin{matrix} CH_3 \quad OH \\ \diagdown \diagup \\ C \\ \diagup \diagdown \\ CH_3 \quad R \end{matrix} \quad + Mg(OH)X$$

Aldehydes yield secondary alcohols as a result of this reaction, while ketones produce tertiary alcohols. Methanal gives rise to primary alcohols.

## Preparation and Uses of Methanol

Methanol is produced:

1 By the distillation of hard wood in the absence of air.
2 From carbon monoxide and hydrogen at 300°C and at high pressures, in the presence of a chromium (III) oxide catalyst which incorporates some zinc oxide:

$$CO + 2H_2 \rightleftharpoons CH_3OH$$

An excess of hydrogen is used in this reaction, and any unchanged carbon monoxide is removed by dissolution in copper (I) methanoate solution. This synthetic process accounts for most of the methanol produced on an industrial scale.

Methanol is used:

(a) For the production of methanal which is used in the manufacture of plastics.
(b) As a solvent for paints, varnishes, oils and polishes.

## Preparation and Uses of Ethanol

Ethanol has been prepared throughout the centuries by the fermentation of starches and sugars. Enzymes, which may be regarded as complex organic catalysts, accelerate the breakdown of starches into sugars, and the enzymes in yeast speed up the conversion of sugars into alcohol. For example, zymase catalyses the breakdown of glucose:

$$C_6H_{12}O_6 \xrightarrow{\text{zymase}} 2CO_2 + 2C_2H_5OH$$

Ethanol produced by fermentation is used in industry, but more especially for consumption in the form of beer, wines and spirits.

Most of the ethanol used in chemical manufacture is made from ethene, originally by the absorption of this gas in sulphuric acid and the subsequent hydrolysis of the products. The direct hydration of ethene has been introduced more recently. A mixture of steam and ethene is passed over a phosphoric acid catalyst at 300°C and at high pressure:

$$CH_2{=}CH_2 + H_2O \rightarrow CH_3CH_2OH$$

Ethanol is used:

(a) For consumption in the form of beer, wines and spirits. Any alcoholic drink is subject to customs tax, which is levied according to the alcoholic content of the drink. Since the specific gravity of ethanol is 0·789, a hydrometer is used to indicate the proportion of alcohol in a water–ethanol mixture, and the results are expressed in terms of degrees under or over proof. Proof spirit has a specific gravity of 0·923.
(b) As a raw material for the manufacture of other chemicals including ethanal, ethanoic acid and esters, etc.
(c) As a solvent.
(d) As a fuel. It is used in small heating stoves, while some brands of petrol contain a small proportion of ethanol.

Ethanol (b.p. 78·4°C) forms an azeotrope with water, containing 95·5 per cent ethanol. This mixture is sold as *rectified spirit*. The water remaining in rectified spirit may be removed by chemical dehydration (using freshly made calcium oxide) or by extractive distillation. In this method, benzene is added to the ethanol–water azeotrope and the mixture is fractionated. A benzene–water–ethanol fraction (b.p. 65°C) is first removed, followed by a fraction containing ethanol and any excess benzene (b.p. 68°C). The final fraction is pure ethanol. Ethanol containing 0·5 per cent water or less is termed *absolute alcohol*.

*Industrial methylated spirit* is rectified spirit which has been denatured by the addition of 5 per cent of methanol, making it unfit to drink. (See below.) Mineralized methylated spirit contains 9 per cent of methanol together with small amounts of pyridine and blue dye which give an unpleasant odour, taste and colour to the liquid.

## Properties and Reactions of the Monohydric Alcohols

The alcohols are colourless liquids having a characteristic odour and burning taste. The lower members are soluble in water, forming a solution which is neutral to litmus, but the miscibility decreases as the relative molecular mass increases. For example, butan-1-ol is only slightly miscible with water, butan-2-ol has an unusual partial miscibility curve, while 2-methylpropan-2-ol is completely soluble in water. Methanol is poisonous. Small quantities of it may bring about a degree of blindness or paralysis, but ethanol can be assimilated in small measures. The reactions of the monohydric alcohols are chiefly those of the functional group—the hydroxyl group. The reactions of ethanol will be given as typical of the series, although the differences in behaviour between

    (a) methanol and ethanol,
    (b) primary, secondary and tertiary alcohols

will be made apparent where this is relevant.

1 The lower alcohols burn in air with a non-luminous flame:

$$CH_3CH_2OH + 3O_2 \rightarrow 2CO_2 + 3H_2O$$

2 Although the alcohols are neutral and do not ionize, metallic sodium reacts with an alcohol to liberate hydrogen and produces a crystalline ionic compound called an alkoxide:

$$2CH_3CH_2OH + 2Na \rightarrow H_2 + 2CH_3CH_2ONa$$
sodium ethoxide

Primary alcohols react more readily than secondary alcohols, which in turn react more readily than tertiary alcohols.

3 Phosphorus trihalides react to give a halogeno-alkane:

$$3CH_3CH_2OH + PX_3 \rightarrow H_3PO_3 + 3CH_3CH_2X$$

Phosphorus pentachloride reacts very readily with the hydroxyl group producing a chloro-alkane:

$$R.CH_2OH + PCl_5 \rightarrow POCl_3 + HCl + R.CH_2Cl$$

4 Alcohols react with hydrogen halides giving halogeno-alkanes:

$$R.CH_2OH + HBr \rightarrow H_2O + R.CH_2Br$$

Tertiary alcohols undergo this reaction more readily than either primary or secondary alcohols.

5 On refluxing an alcohol with a carboxylic acid, an ester and water are formed. The reaction is reversible, and is catalysed by hydrogen ion. Hence, a small quantity of concentrated sulphuric acid is added to the reaction mixture. For example:

$$CH_3CH_2OH + CH_3CO_2H \rightleftharpoons CH_3CO_2CH_2CH_3 + H_2O$$
ethyl ethanoate

Esters made from 3,5-dinitrobenzoic acid,

are solid, and have different melting points according to the alcohol used. These derivatives are used to identify the individual alcohols.

6 Passing alcohol vapour over hot activated alumina results in the formation of an alkene by dehydration:

$$CH_3CH_2OH - H_2O \rightarrow CH_2{=}CH_2$$

(Methanol cannot react in this way.) Tertiary alcohols dehydrate most readily, while primary alcohols are the least reactive.

Alternatively, dehydration may be effected by the use of hot syrupy phosphoric acid, or by hot concentrated sulphuric acid:

$$CH_3CH_2OH + H_2SO_4 \rightarrow H_2O + CH_3CH_2SO_4H$$
ethyl hydrogen sulphate

On heating further (keeping the sulphuric acid in excess), ethene is produced:

$$CH_3CH_2SO_4H \rightarrow H_2SO_4 + CH_2{=}CH_2$$

However, if more alcohol is added, ethoxy-ethane is produced:

$$CH_3CH_2SO_4H + CH_3CH_2OH$$
$$\rightarrow H_2SO_4 + CH_3CH_2\text{---}O\text{---}CH_2CH_3$$

7 Alcohols may be oxidized to carbonyl compounds or to carb-oxylic acids. The actual product of this reaction depends on the type of alcohol involved, and the results of oxidation are useful in classifying alcohols.

Primary alcohols yield an aldehyde, which is easily oxidized further producing a carboxylic acid:

$$R.CH_2OH \xrightarrow{O} R.CHO \xrightarrow{O} R.CO_2H$$

Secondary alcohols produce ketones:

$$\begin{array}{c} R \\ \phantom{R'}\diagdown \\ \phantom{R}CHOH \\ \diagup \\ R' \end{array} \xrightarrow{O} \begin{array}{c} R \\ \phantom{R'}\diagdown \\ \phantom{R}C=O \\ \diagup \\ R' \end{array}$$

Ketones are more resistant to further oxidation.

Aldehydes and ketones (p. 79) are readily distinguished by simple chemical tests, so that primary and secondary alcohols can be classified on the results of this reaction. Tertiary alcohols are not easily oxidized, but under drastic conditions they break down to give products such as carbon dioxide, ketones and acids, all containing fewer carbon atoms than the original alcohol. The most commonly used oxidizing agent is potassium dichromate in dilute sulphuric acid solution, although methanol and ethanol are oxidized by air when the alcohol vapour is passed over a hot copper or platinum gauze:

$$2CH_3OH + O_2 \rightarrow 2H_2O + 2HCHO$$

8 Alcohols may be reduced to alkanes by heating under pressure with red phosphorus and concentrated hydriodic acid:

$$CH_3CH_2OH + 2HI \rightarrow CH_3CH_3 + I_2 + H_2O$$

The liberated iodine is converted back to hydrogen iodide by the red phosphorus.

9 *The haloform reaction.* Ethanol reacts with chlorine in alkaline solution to produce trichloromethane. The stages are given on p. 48.

$$CH_3CH_2OH + 4Cl_2 + NaOH$$
$$\rightarrow CHCl_3 + 5HCl + HCOONa$$

A similar reaction takes place with bromine and iodine. In the latter reaction, the product, tri-iodomethane (iodoform), is a pale yellow crystalline solid. During this sequence of reactions, a carbon atom is lost, and it follows that methanol cannot undergo this reaction. This test may therefore be used to distinguish methanol from ethanol.

TABLE 8 *The differences between methanol and ethanol*

|  | Methanol | Ethanol |
|---|---|---|
| *Haloform reaction using iodine* |  | crystals of tri-iodomethane produced |
| *Reaction with concentrated sulphuric acid* | forms dimethyl sulphate $(CH_3)_2SO_4$ | ethyl hydrogen sulphate is formed first; on heating either ethene or ethoxy-ethane is produced according to the conditions |
| *Oxidation* | methanal produced | ethanal formed |
| *Reduction with phosphorus and HI* | methane formed | ethane produced |
| *m.p. of 3,5-dinitro-benzoate* | 107°C | 93°C |

## Names of the Alcohols

The common names of the alcohols are derived from the name of the alkyl group present in the molecule, and these are shown in Table 7 on p. 52. The systematic name consists of four parts:

1 The number of carbon atoms in the longest straight chain determines the prefix as given on p. 32.
2 The second part of the name shows whether the alcohol is saturated (in which case -an is used) or contains a double bond (when -en is used).
3 A number indicates the carbon atom which is linked to the hydroxyl group.
4 The name finally ends in -ol to show that a hydroxyl group is present.

Thus,

$$CH_3CH_2CH_2CHCH_3$$
$$|$$
$$OH$$

is pentan-2-ol, to agree with the rule that the numbers should be kept as small as possible.

## The Structural Formula of Ethanol

1 Qualitative analysis shows ethanol to contain the elements carbon, hydrogen and possibly oxygen only.

2 Quantitative analysis gives 52·17 per cent carbon, 13·04 per cent hydrogen, and, by difference, 34·79 per cent oxygen. Dividing these through by the respective relative atomic masses,

$$C = \frac{52·17}{12·01} = 4·347 \qquad H = \frac{13·04}{1·008} = 12·91$$

$$O = \frac{34·79}{16·00} = 2·167$$

Dividing these numbers by the smallest,

$$C = \frac{4·347}{2·167} = 2·01 \qquad H = \frac{12·91}{2·167} = 5·98$$

$$O = \frac{2·167}{2·167} = 1·00$$

giving an empirical formula of $C_2H_6O$.

3 The vapour density of ethanol is 23, and so the relative molecular mass is 46. Hence the molecular formula is $C_2H_6O$.

4 Taking carbon, hydrogen and oxygen as having covalencies of four, one and two respectively, two possible alternative structures for ethanol may be suggested:

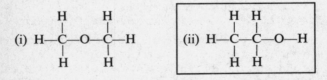

Further evidence is now required before the correct structure can be deduced.

(a) Ethanol reacts with sodium to form the ethoxide, which is shown by analysis to have the formula $C_2H_5ONa$. By experiment, two moles of ethanol yield one mole of hydrogen during this reaction, so that each molecule of ethanol loses one atom of hydrogen. No further hydrogen can be displaced from ethanol or sodium ethoxide in this way. This suggests that in ethanol one atom of hydrogen is linked in a different way from the other five, i.e. $C_2H_5O(H)$.

(b) Phosphorus pentachloride (or trichloride) reacts readily with ethanol to produce a gas of formula $C_2H_5Cl$. This gas is identical with that produced by the substitution of one hydrogen atom in ethane by chlorine; the gas is chloroethane. Although chloroethane reacts with sodium, no hydrogen is liberated. Hence the hydrogen in ethanol which reacts with sodium has also been lost during the reaction with phosphorus pentachloride. This suggests that the formula of ethanol is $C_2H_5(OH)$.

(c) Ethanol may be produced by chlorinating ethane to form chloroethane and reacting this with alkali:

$$CH_3CH_3 \xrightarrow{Cl_2} CH_3CH_2Cl \xrightarrow{NaOH} CH_3CH_2OH$$

In view of this evidence, structure (ii) is taken to represent the structural formula of ethanol.

# 6 Ethers

Ethers have the general formula R—O—R′, where R and R′ are alkyl groups linked by a bridging oxygen atom.

TABLE 9 *Lower ethers*

| Name | Formula | B.P. °C |
|---|---|---|
| methoxy-methane | $CH_3$—O—$CH_3$ | −24·1 |
| ethoxy-ethane (ether) | $CH_3CH_2$—O—$CH_2CH_3$ | 34·6 |
| methoxy-ethane | $CH_3$—O—$CH_2CH_3$ | 9·3 |

Simple ethers are those in which the two alkyl groups are identical, while a mixed ether contains two different alkyl groups.

## General Methods of Preparation

1 By the partial dehydration of the corresponding alcohol (p. 58):

$$2R.OH - H_2O \rightarrow R—O—R$$

Alcohol is dehydrated using concentrated sulphuric acid. The ether and water may be removed by distillation as fast as fresh alcohol is added. This process is known as *Williamson's continuous etherification process*. It is capable of producing simple ethers only.

2 *Williamson's synthesis* This process involves the reaction of an alkyl halide with sodium alkoxide:

$$R.Br + R'.ONa \rightarrow R—O—R' + NaBr$$

Both simple and mixed ethers may be produced by this reaction.

## Properties and Reactions of the Ethers

The lower members of the series are volatile, sweet smelling liquids (methoxy-methane is a gas) which are slightly soluble in water.

The ethers are not particularly reactive. The reactions of ethoxy-ethane (usually known as ether) are given as typical of the series.

1 They burn readily in air:

$$C_2H_5—O—C_2H_5 + 6O_2 \rightarrow 4CO_2 + 5H_2O$$

A mixture of ether vapour and air is highly inflammable and often explosive. Ethoxy-ethane, commonly used as an organic solvent, carries a high fire risk, since

(a) the liquid has a low boiling point and is very volatile,
(b) ether vapour is heavier than air, and thus it is not dispersed in an ascending air stream, but accumulates in low-lying pockets such as sinks and drains, etc.

2 Another danger associated with ethers is the slow formation of a small quantity of a peroxide (R—O—O—R′) when ethers stand in contact with oxygen or air. Peroxides are unstable compounds and explode on heating. They are removed on shaking with a little iron (II) sulphate.

3 Hydriodic acid slowly reacts with ethers:

$$C_2H_5—O—C_2H_5 + HI \rightarrow C_2H_5OH + C_2H_5I$$

If a large excess of hydriodic acid is present, two molecules of iodoethane are produced:

$$C_2H_5—O—C_2H_5 + 2HI \rightarrow H_2O + 2C_2H_5I$$

4 Ethers combine with strong mineral acids at low temperatures giving products known as *oxonium salts*.

$$\begin{matrix} C_2H_5 \\ \diagdown \\ \diagup \\ C_2H_5 \end{matrix} O + HCl \rightarrow \left[ \begin{matrix} C_2H_5 \\ \diagdown \\ \diagup \\ C_2H_5 \end{matrix} O \rightarrow H \right]^{+} Cl^{-}$$

A coordinate link joins the oxygen atom in the ether to the proton of the acid.

**Uses** Ethoxy-ethane is used as a solvent and as an anaesthetic.


**Isomerism**

Ethers provide an example of a type of isomerism known as *metamerism*. The isomers (metamers) belong to the same homologous series, and their differences result from the presence of different pairs of alkyl groups linked to the same functional group in the molecule. For example, ethoxy-ethane, $C_2H_5OC_2H_5$ (b.p. 34·6°C), and methoxy-propane, $CH_3OCH_2CH_2CH_3$ (b.p. 39°C), are metamers.

In addition, every ether is isomeric with an alcohol (e.g. $CH_3CH_2OH$ and $CH_3OCH_3$).

63

# 7  Carbonyl Compounds

## ALDEHYDES

The general formula for the homologous series of aldehydes is

$$\begin{array}{c} R \\ \diagdown \\ C=O \\ \diagup \\ H \end{array}$$

where R represents an alkyl group. The functional group is the carbonyl group $\diagup\!\!\!C=O$, but the single hydrogen atom attached to this group also displays a special reactivity.

TABLE 10 *Lower members of the aldehyde series*

| Name | Formula | M.P. °C | B.P. °C |
|------|---------|---------|---------|
| methanal (formaldehyde) | HCHO | − 92 | −21 |
| ethanal (acetaldehyde) | $CH_3CHO$ | −123 | 20·8 |
| propanal | $CH_3CH_2CHO$ | − 81 | 48·8 |
| butanal | $CH_3CH_2CH_2CHO$ | − 97 | 74·7 |

The IUPAC system of nomenclature demands the ending -al with the suffix assigned according to the number of carbon atoms in the molecule as given in Table 3 on p. 32.

## General Methods of Preparation

1 By the oxidation of a primary alcohol, using an oxidizing agent such as acidified potassium dichromate:

$$R.CH_2OH + (O) \rightarrow R.CHO + H_2O$$

As aldehydes are very readily oxidized to carboxylic acids, it is necessary to use a preparative technique which cuts down the loss of product in this way. The aldehydes produced by the oxidation of a primary alcohol boil at a lower temperature than the parent alcohol—for example, ethanol boils at 78·4°C while ethanal boils at 20·8°C. Consequently, ethanol is added drop by drop to the oxidizing solution which is kept above the boiling point of ethanal but below that of ethanol, at 50°C for example. As the ethanal is produced, it is immediately vaporized and removed from the oxidizing medium. Even so, a significant quantity of ethanoic acid does result from this reaction.

2 By heating a mixture of calcium methanoate and the calcium salt of the appropriate acid:

$$(RCO_2)_2Ca + (HCO_2)_2Ca \rightarrow 2CaCO_3 + 2RCHO$$

The yield in most cases is low, and methanal and a ketone are formed as side products.

3 By the hydrogenation of an acid chloride. This is known as the *Rosenmund* reaction.

$$R.COCl + H_2 \xrightarrow{\text{catalyst}} R.CHO + HCl$$

The catalyst is of particular importance since it has to speed up the replacement of the chlorine in the acid chloride by hydrogen, and yet must not favour the reduction of the aldehyde. Rosenmund discovered that a palladium catalyst whose efficiency was impaired (i.e. a poisoned catalyst) by the presence of sulphur compounds (especially barium sulphate) was particularly suitable in this reaction.

## Special Methods of Preparation

1 *Methanal* is produced by the oxidation of methanol by passing a mixture of air and methanol vapour over a copper or silver gauze. In the manufacture of methanal, the oxidation takes place with the aid of a catalyst which consists of silver granules or a mixture of metal oxides:

$$2CH_3OH + O_2 \rightarrow 2HCHO + 2H_2O$$

2 *Ethanal* may be produced by the hydration of ethyne (p. 43):

$$CH{\equiv}CH + H_2O \rightarrow CH_3CHO$$

In view of the high cost of ethyne, the production of ethanal by this method has been abandoned in favour of the following cheaper processes.

(a) Ethanol vapour is passed over a hot silver catalyst. Oxidation and dehydrogenation take place simultaneously:

$$2CH_3CH_2OH + O_2 \rightarrow 2CH_3CHO + 2H_2O$$

$$2CH_3CH_2OH \rightarrow 2CH_3CHO + H_2$$

(b) Ethene and oxygen are fed into an acidic solution containing palladium (II) chloride and copper (II) chloride at 50°C:

$$CH_2\!=\!CH_2 + H_2O + 2CuCl_2 \rightarrow CH_3CHO + 2HCl + 2CuCl$$

Then

$$4CuCl + 4HCl + O_2 \rightarrow 4CuCl_2 + 2H_2O$$

and thus the cycle continues.


## Properties and Reactions of the Aldehydes

Methanal is a gas at room temperature; ethanal and the lower aldehydes are liquids. The aldehydes have characteristic choking odours, and although the lower members are very soluble in water, the solubility decreases as the relative molecular mass increases.

The main reactions of the aldehydes can be discussed under three headings:

(a) reactions of the alkyl group,

(b) oxidation of the $-\overset{\shortparallel}{\underset{O}{C}}-H$ group to $-\overset{\shortparallel}{\underset{O}{C}}-O-H$,

(c) reactions of the carbonyl group $\diagdown C\!=\!O$, such as addition, condensation and polymerization.

The carbonyl group is present in both aldehydes and ketones, and they show many reactions in common. For aldehydes, the $-CHO$ group (the aldehyde group) is often quoted as the functional group. The reactions of ethanal are given as typical of the series:

1 Aldehydes burn in air:

$$2CH_3CHO + 5O_2 \rightarrow 4CO_2 + 4H_2O$$

2 Substitution of the hydrogen atoms in the alkyl group by a halogen:

$$CH_3CHO + 3Cl_2 \rightarrow 3HCl + CCl_3CHO$$

The product, trichloroethanal (or chloral) may be hydrolysed to give trichloromethane, or chloroform (p. 48). Bromine and iodine react in a similar manner.

Methanal has no alkyl group, and cannot take part in this type of reaction.

3 *Oxidation* of the —CHO group to —CO₂H: this reaction may be brought about by mild oxidizing agents; conversely, aldehydes act as reducing agents. For example:

(a) Ammoniacal silver nitrate is reduced to silver:

$$2Ag^+ + CH_3CHO + H_2O \rightarrow CH_3CO_2^- + 3H^+ + 2Ag \downarrow$$

The deposition of a film of silver on the surface of the vessel or tube in which the reaction is being carried out leads to this test being described as the *silver mirror test*. Tollen's reagent, a solution of $Ag^+$ in alkali, reacts in a similar way.

(b) Fehling's solution—an alkaline solution of a copper (II) salt, stabilized by the addition of tartrate ion—is reduced to copper (I) by aldehydes on gentle heating:

$$2Cu^{2+} + CH_3CHO + H_2O \rightarrow CH_3CO_2^- + 3H^+ + 2Cu^+ \downarrow$$

The blue colour of the Fehling's solution changes to green, yellowish and finally orange corresponding to the formation of orange-red copper (I) oxide.

(c) Potassium dichromate is reduced to chromium (III) sulphate:

$$3CH_3CHO + K_2Cr_2O_7 + 4H_2SO_4$$
$$\rightarrow 3CH_3CO_2H + 4H_2O + K_2SO_4 + Cr_2(SO_4)_3$$

As a result of this reaction, potassium and chromium (III) sulphates are formed in equimolar proportions; hence the reaction is used for making the alum $KCr(SO_4)_2 . 12H_2O$.

*Reactions of the carbonyl group* are as follows (4–7).

4 *Reduction*, the products depending on the conditions of the reaction:

(a) Reduction of ethanal using sodium amalgam, lithium aluminium hydride or zinc and dilute acid yields the corresponding alcohol:

$$CH_3CHO + H_2 \rightarrow CH_3CH_2OH$$

(b) Reduction using amalgamated zinc and concentrated hydrochloric acid produces ethane:

$$CH_3CHO + 2H_2 \rightarrow CH_3CH_3 + H_2O$$

This reaction is known as the *Clemmensen reduction*.

5 *Addition* to the double bond of the carbonyl group. (Addition reactions are those in which two or more compounds unite to give a single product.)

(a) When well shaken with saturated sodium hydrogen sulphite solution, a crystalline derivative is formed:

$$CH_3C\begin{smallmatrix}H\\ \\O\end{smallmatrix} + NaHSO_3 \rightarrow CH_3-\underset{H}{\overset{OH}{C}}-O.SO_2^- Na^+$$

The formation of these addition compounds is reversible, and the original aldehyde may be regenerated using sodium carbonate or dilute acid. The formation of these crystalline compounds affords a method of separating and purifying aldehydes.

(b) Hydrogen cyanide also adds on to aldehydes. Ethanal is mixed with an aqueous solution of potassium cyanide and dilute acid is added:

$$CH_3C\begin{smallmatrix}H\\ \\O\end{smallmatrix} + HCN \rightarrow CH_3-\underset{H}{\overset{OH}{C}}-CN$$

By means of this reaction, an extra carbon atom is introduced into the molecule.

(c) Aldehydes react with dry ammonia gas to produce a white precipitate of an aldehyde ammonia:

$$R-C\begin{smallmatrix}H\\ \\O\end{smallmatrix} + NH_3 \rightarrow R-\underset{NH_2}{\overset{H}{C}}-OH$$

Other, more complex products are formed during this reaction. Aldehyde ammonias are not very stable, and are decomposed to give the original aldehyde on treating with acid.

Methanal reacts in a different manner with ammonia. On the evaporation of a solution of methanal and concentrated ammonia, a white solid, hexamethylene tetramine (or hexamine) is formed. This compound has a complex cyclic structure.

$$6HCHO + 4NH_3 \rightarrow \quad [\text{hexamethylenetetramine structure}] \quad + 6H_2O$$

6 *Polymerization.* In common with many unsaturated organic molecules, aldehydes undergo polymerization reactions. The products vary according to the conditions and reagents required to bring about this reaction.

(a) When gaseous methanal, produced by boiling a solution of methanal containing a little sulphuric acid is kept at room temperature, a white solid trimer (i.e. a polymer resulting from the union of three monomer molecules) is formed:

$$3HCHO \rightarrow \quad [\text{trioxymethylene ring structure}]$$

The polymer is trioxymethylene, and it decomposes on heating to give pure, dry methanal gas. It is thus a convenient way of storing methanal. Under normal conditions, methanal is a water-soluble gas, and a more common method of storing it is in the form of a solution (known as formalin) which contains 38 per cent by weight of methanal.

(b) Ethanal undergoes rapid polymerization in the presence of a trace of concentrated sulphuric acid to form the trimer, paraldehyde.

$$3CH_3CHO \rightarrow \quad [\text{paraldehyde ring structure}]$$

Paraldehyde is a liquid which does not reduce Fehling's

solution. On adding dilute acid, ethanal is regenerated so that paraldehyde gives the reactions of ethanal when tested in acid solution.

(c) A second solid polymer of ethanal known as metaldehyde can be made by passing dry hydrogen chloride through a solution of ethanal in ether, while the temperature is kept below 0°C. Metaldehyde has the formula $(CH_3CHO)_4$.

(d) On boiling ethanal with dilute sodium hydroxide solution, polymerization to a brown resin takes place. In similar circumstances, methanal reacts in a different way, undergoing the *Cannizzaro reaction*. This reaction may be regarded as the oxidation of one molecule of methanal to methanoic acid by another molecule of methanal, which is, in turn, reduced to methanol.

$$2HCHO + NaOH \rightarrow CH_3OH + HCO_2Na$$

This reaction is useful in differentiating between methanal and ethanal.

7 *Condensation reactions.* A condensation reaction is one in which two molecules react to give a product molecule and water only,† i.e. $A + B \rightarrow$ product $+ H_2O$. (The opposite of this reaction is hydrolysis.) With aldehydes (and ketones) the oxygen of the carbonyl group is eliminated by condensation.

(a) On reacting with hydroxylamine, the condensation product is called an oxime:

$$CH_3CHO + H_2NOH \rightarrow H_2O + CH_3C\begin{smallmatrix}H \\ \| \\ N-OH\end{smallmatrix}$$
hydroxylamine

(b) With phenylhydrazine, the condensation product is a phenyl-hydrazone:

$$CH_3CHO + H_2NNHC_6H_5 \rightarrow H_2O + CH_3C\begin{smallmatrix}H \\ \| \\ NNHC_6H_5\end{smallmatrix}$$
phenylhydrazine

ethanal phenylhydrazone

Condensation reactions usually take place readily on mixing the two reactants in a suitable solvent. When 2,4-dinitro-phenylhydrazine* is used, the resulting condensation products

† Alternatively a condensation reaction is one in which two molecules react to give a product molecule and water, or another small molecule only.

* A solution of 2,4-dinitrophenylhydrazine is known as 'Bradys' reagent.

are solids with sharp melting points. In this way, individual aldehydes (and ketones) may be readily identified.

(c) Semicarbazones are condensation products resulting from the reaction of an aldehyde with semicarbazide:

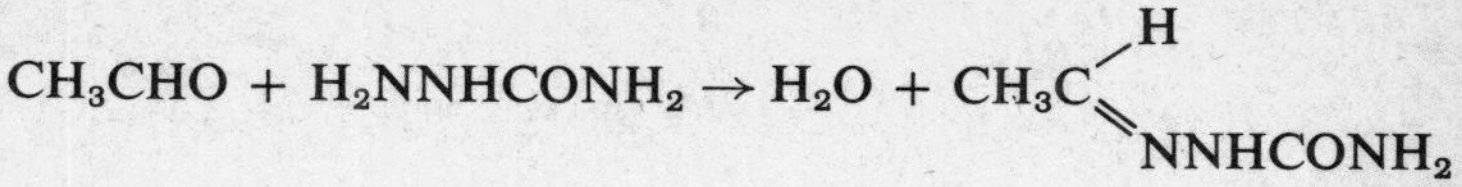

$$CH_3CHO + H_2NNHCONH_2 \rightarrow H_2O + CH_3C{\overset{\textstyle H}{\underset{\textstyle NNHCONH_2}{\big\Vert}}}$$

Other reactions of aldehydes include:

8 When a solution of ethanal is mixed with a *small* quantity of dilute alkali (i.e. sodium carbonate or hydroxide) a reaction known as the *aldol condensation* takes place:

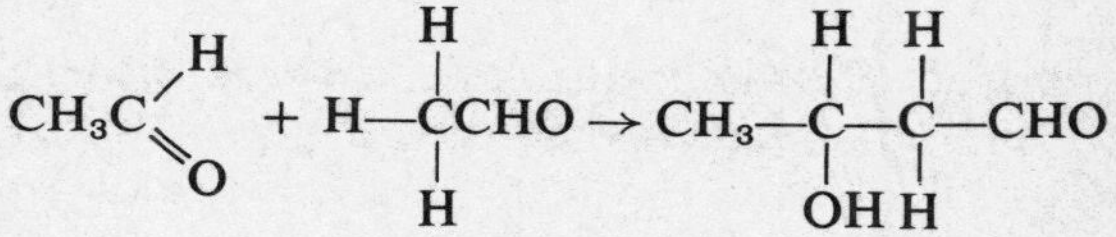

$$CH_3C{\overset{\textstyle H}{\underset{\textstyle O}{\big\Vert}}} + H{-}\underset{\textstyle H}{\overset{\textstyle H}{\underset{|}{\overset{|}{C}}}}CHO \rightarrow CH_3{-}\underset{\textstyle OH}{\overset{\textstyle H}{\underset{|}{\overset{|}{C}}}}{-}\underset{\textstyle H}{\overset{\textstyle H}{\underset{|}{\overset{|}{C}}}}{-}CHO$$

The product (aldol or 3-hydroxybutanal) has both an *ald*ehyde and an alco*hol* group and is a polymer (dimer) rather than a condensation product.

9 Aldehydes produce a colour with *Schiff's reagent*. This reagent is a solution of a magenta dye which has been decolorized by passing sulphur dioxide through the solution.

10 Phosphorus pentachloride reacts with aldehydes by displacing the oxygen atom present in the carbonyl group. No hydrogen chloride is evolved, in contrast to the reaction between phosphorus pentachloride and a hydroxyl group.

$$CH_3CHO + PCl_5 \rightarrow CH_3CHCl_2 + POCl_3$$

Phosphorus oxychloride and 1,1-dichloroethane are formed.

**Uses of methanal**  Aqueous solutions of methanal are used as antiseptics and preservatives. Some methanal is used to manufacture hexamine (p. 68), which when nitrated produces a powerful explosive (cyclonite or RDX). The major use of methanal is in the manufacture of plastics by condensation polymerization. This type of polymerization results from the repeated union of two molecules by the elimination of the elements of water between them. A typical example is the formation of urea-formaldehyde polymer:

$$H_2N-\underset{\underset{O}{\|}}{C}-N\underset{H}{\overset{H}{<}} \quad + \quad \underset{H}{\overset{\overset{\displaystyle O}{\|}}{\underset{\displaystyle}{C}}}\underset{H}{} \quad + \quad \overset{H}{\underset{H}{>}}N-\underset{\underset{O}{\|}}{C}-NH_2$$

urea

$$\rightarrow H_2O + H_2N-\underset{\underset{O}{\|}}{C}-\underset{\underset{H}{|}}{N}-\underset{\underset{H}{|}}{\overset{\overset{H}{|}}{C}}-\underset{\underset{H}{|}}{N}-\underset{\underset{O}{\|}}{C}-NH_2$$

$$\left( -\underset{\underset{H}{|}}{N}-\underset{\underset{O}{\|}}{C}-\underset{\underset{H}{|}}{N}-\underset{\underset{H}{|}}{\overset{\overset{H}{|}}{C}}- \right)_n \quad \leftarrow \quad \text{continuing to give a polymer containing the repeating unit}$$

**Uses of ethanal**  Ethanal is manufactured mainly for conversion to ethanoic acid, although this process is gradually being replaced (p. 81). It is also used as an intermediate in the manufacture of a limited range of plastics, resins and drugs.

### Theory of the Addition Reactions of Carbonyl Compounds

In these compounds, the carbon and oxygen atoms forming the carbonyl group are linked by a double bond, and, as expected, addition reactions take place. On comparing these reactions with the addition reactions of the alkenes, it is clear that the reagents which add on to the two classes of compound are different. For example, chlorine adds on to an alkene, but substitutes the alkyl hydrogen atoms in aldehydes and ketones.

The addition reactions of aldehydes and ketones are catalysed by acids and bases, showing that heterolytic fission is involved. The electronegative oxygen atom in the carbonyl group pulls electrons away from the carbon to produce a partial positive charge on that carbon. This carbon atom now becomes subject to attack by a nucleophile and the initial step in the mechanism is the formation of a link between the nucleophile and the carbon atom. The oxygen, now carrying a full negative charge, links with the electrophile; the sequence being shown in Figure 9.

### Isomerism

Apart from methanal and ethanal, every aldehyde is isomeric with at least one ketone; propanal $CH_3CH_2CHO$ with propanone

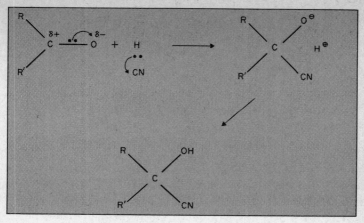

Fig. 9 Addition to the carbonyl group

$CH_3COCH_3$, for example. The usual straight- and branched-chain isomers are found in the alkyl group.

A further point of interest is that oximes are isomeric with amides; for example,

$$CH_3C\overset{H}{\underset{NOH}{\diagup}} \qquad \text{and} \qquad CH_3CONH_2$$

### The Structural Formula of Ethanal

1 Qualitative and quantitative analysis show ethanal to have the empirical formula $C_2H_4O$.

2 Relative molecular mass determinations show that this is also the molecular formula.

3 Sodium does not displace hydrogen in the cold from ethanal, and the reaction between phosphorus pentachloride and ethanal does not yield hydrogen chloride. Both these observations suggest the absence of a hydroxyl group.

4 Only three hydrogen atoms are replaced on the chlorination of ethanal.

5 Ethanal is prepared by the oxidation of ethanol, so that possible structures of ethanal are:

73

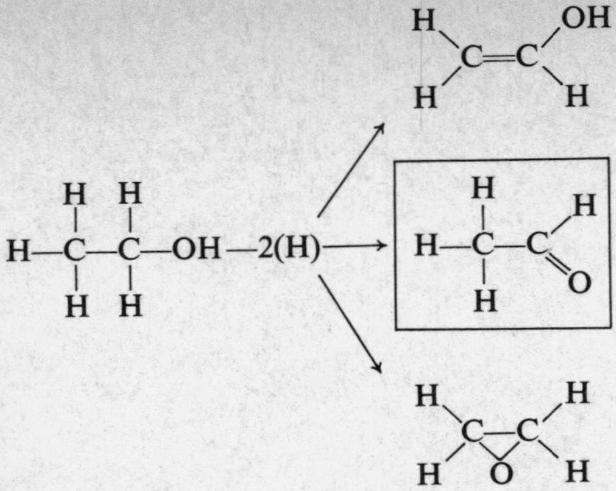

(a)

(b)

(c)

6 Structure (a) is eliminated as it contains a hydroxyl group, and (c) is rejected since it contains four equivalent hydrogen atoms which should be substituted on chlorination. Structure (c) is also rejected since reaction with phosphorus pentachloride should produce 1,2-dichloroethane.

7 Physical methods, such as infrared and nuclear magnetic resonance spectroscopy, indicate (i) the presence of a double bond linking the carbon and oxygen atoms, and (ii) that three of the four hydrogen atoms are located in equivalent positions in the molecule.

In view of this evidence, structure (b) is the structural formula assigned to ethanal.

## KETONES

The general formula for the ketones is $\begin{array}{c} R \\ \diagdown \\ \diagup \\ R' \end{array} C{=}O$ where R and R' represent alkyl groups (which may be the same or different). The functional group is the carbonyl group.

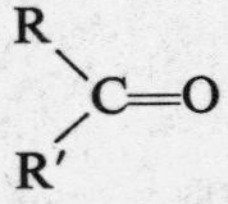

The IUPAC system uses the ending -one in connection with the suffixes given in Table 3 on p. 32. The number inserted in this name shows the number of the carbon atom which forms part of the carbonyl group.

TABLE 11 *The lower members of the ketone series*

| Name | Formula | M.P.°C | B.P.°C |
|---|---|---|---|
| propanone (dimethyl ketone or acetone) | $CH_3CCH_3$ $\parallel$ O | $-95$ | 56·1 |
| butan-2-one | $CH_3CCH_2CH_3$ $\parallel$ O | $-85$ | 79·6 |
| pentan-3-one | $CH_3CH_2CCH_2CH_2$ $\parallel$ O | $-42$ | 101·6 |

## Isomerism

Ketones show metamerism. Each ketone has at least one isomeric aldehyde, and isomerism due to branching in the alkyl group is possible, as with pentan-2-one and 3-methylbutan-2-one:

$$CH_3—C—CH_2—CH_2—CH_3 \qquad CH_3—CH—C—CH_3$$
$$\underset{O}{\parallel} \qquad\qquad\qquad\qquad \underset{CH_3\ \ O}{|\quad\ \parallel}$$

pentan-2-one                3-methylbutan-2-one

## General Methods of Preparation

1 By the oxidation of a secondary alcohol:

$$\begin{array}{c} R \\ \diagdown \\ \diagup \\ R' \end{array}\!\!CHOH + (O) \rightarrow \begin{array}{c} R \\ \diagdown \\ \diagup \\ R' \end{array}\!\!C{=}O + H_2O$$

Potassium dichromate dissolved in dilute sulphuric acid is the usual oxidizing agent. Unlike aldehydes, ketones are not easily oxidized; therefore no special technique is necessary to prevent further oxidation to an acid.

2 By heating the calcium salt of an acid:

$$(R.CO_2)_2Ca \rightarrow \begin{array}{c} R \\ \diagdown \\ \diagup \\ R \end{array}\!\!C{=}O + CaCO_3$$

The yield of ketone is often low, and mixed ketones (i.e. those which contain different alkyl groups) cannot be made by this method because of the spread of products obtained.

75

**Manufacture**  Propanone is manufactured by the dehydrogenation of propan-2-ol, by passing the vapour over a mixture of copper and zinc oxides at 350°–400°C:

$$\begin{array}{c} CH_3 \\ \diagdown \\ CH_3 \end{array} CHOH \rightarrow H_2 + \begin{array}{c} CH_3 \\ \diagdown \\ CH_3 \end{array} C{=}O$$

## Properties and Reactions of the Ketones

The ketones are pleasant smelling, colourless liquids (those with a high relative molecular mass are solid at room temperature) which are generally insoluble in water. Propanone, however, is freely miscible with water.

The main reactions of the ketones are those associated with the carbonyl group; the behaviour of propanone is given as typical of the series. Ketones do not possess an activated hydrogen atom in the molecule, so that, unlike aldehydes, they show no reducing properties.

1 Ketones burn readily in air:

$$CH_3COCH_3 + 4O_2 \rightarrow 3CO_2 + 3H_2O$$

2 Oxidation of ketones takes place under drastic conditions (e.g. potassium dichromate and fairly concentrated sulphuric acid) to produce a mixture of acids (and sometimes carbon dioxide) whose molecules contain fewer carbon atoms than the original ketone:

$$\begin{array}{c} R \\ \diagdown \\ R'CH_2 \end{array} C{=}O + 3(O) \rightarrow R.CO_2H + R'.CO_2H$$

3 *Reactions of the alkyl group.* The hydrogen atoms in the methyl group in propanone are substituted by bubbling chlorine through propanone:

$$CH_3COCH_3 + 6Cl_2 \rightarrow \begin{array}{c} CCl_3 \\ \diagdown \\ CCl_3 \end{array} C{=}O + 6HCl$$

Ketones containing a methyl group attached to a carbonyl group ($CH_3CO-$) undergo the haloform reaction (p. 48) to produce the tri-substituted derivative:

$$\begin{array}{c} CH_3 \\ \diagdown \\ \diagup \\ CH_3 \end{array} C{=}O + 3X_2 \rightarrow \begin{array}{c} CH_3 \\ \diagdown \\ \diagup \\ CX_3 \end{array} C{=}O + 3HX \qquad \text{where X is chlorine,} \\ \text{bromine or iodine.}$$

Hydrolysis of the product gives trichloromethane (chloroform) or tri-iodomethane (iodoform) where X is chlorine or iodine respectively.

*Reactions of the carbonyl group* are as follows (4–8).

4 Reduction of ketones by zinc and dilute acid, or with lithium aluminium hydride, produces a secondary alcohol:

$$\begin{array}{c} CH_3 \\ \diagdown \\ \diagup \\ CH_3 \end{array} C{=}O + H_2 \rightarrow \begin{array}{c} CH_3 \\ \diagdown \\ \diagup \\ CH_3 \end{array} CHOH$$

Using zinc amalgam and concentrated hydrochloric acid (Clemmensen reduction) ketones are reduced to alkanes:

$$CH_3COCH_3 + 2H_2 \rightarrow CH_3CH_2CH_3 + H_2O$$

5 Ketones show similar addition reactions to those given by the aldehydes, and the mechanism of these is parallel to that described on p. 72.

(a) With saturated sodium hydrogen sulphite solution, crystalline addition compounds are formed:

$$\begin{array}{c} R \\ \diagdown \\ \diagup \\ R' \end{array} C{=}O + NaHSO_3 \rightarrow \begin{array}{c} R \quad OH \\ \diagdown \diagup \\ C \\ \diagup \diagdown \\ R' \quad SO_3.Na \end{array}$$

Since the parent ketone may be regenerated on acidification, the purification of ketones can be accomplished by means of compound formation.

(b) Hydrogen cyanide adds on to ketones:

$$\begin{array}{c} CH_3 \\ \diagdown \\ \diagup \\ CH_3 \end{array} C{=}O + HCN \rightarrow \begin{array}{c} CH_3 \quad OH \\ \diagdown \diagup \\ C \\ \diagup \diagdown \\ CH_3 \quad CN \end{array}$$

(c) With ammonia, a complex series of products is formed, in contrast to the addition reactions shown by the aldehydes.

77

6 Ketones do not polymerize easily, again in contrast to aldehydes. However, in the presence of solid barium hydroxide as a catalyst, two molecules of propanone undergo addition:

$$\begin{array}{c} CH_3 \\ \diagdown \\ C=O \\ \diagup \\ CH_3 \end{array} + CH_3CCH_3 \rightleftharpoons CH_3-\underset{\underset{OH}{|}}{\overset{\overset{CH_3}{|}}{C}}-CH_2-\underset{\underset{O}{\|}}{C}-CH_3$$

This is an example of an aldol type of condensation.

7 Ketones undergo condensation reactions similar to those of the aldehydes.

(a) With hydroxylamine, an oxime is formed:

$$\begin{array}{c} CH_3 \\ \diagdown \\ C=O \\ \diagup \\ CH_3 \end{array} + H_2NOH \rightarrow \begin{array}{c} CH_3 \\ \diagdown \\ C=NOH \\ \diagup \\ CH_3 \end{array} + H_2O$$

(b) Phenylhydrazine reacts to form phenylhydrazones:

$$\begin{array}{c} CH_3 \\ \diagdown \\ C=O \\ \diagup \\ CH_3 \end{array} + H_2NNHC_6H_5 \rightarrow \begin{array}{c} CH_3 \\ \diagdown \\ C=NNHC_6H_5 \\ \diagup \\ CH_3 \end{array} + H_2O$$

propanone phenylhydrazone

The product formed on adding a few drops of a ketone to an alcoholic solution of 2,4-dinitrophenylhydrazine is a yellow solid of characteristic melting point. This reaction is widely used to identify individual ketones.

(c) A special condensation reaction of propanone occurs when it is distilled from concentrated sulphuric acid. Three molecules of propanone combine with the elimination of three molecules of water:

$$3CH_3COCH_3 \rightarrow 3H_2O + \text{[benzene ring with } CH_3 \text{ at top, } CH_3 \text{ lower left, } CH_3 \text{ lower right]}$$

The product, trimethylbenzene, is an example of an aromatic compound being formed from an aliphatic source.

8 The other reactions of ketones are as follows.

(a) With the exception of propanone, ketones do not produce a colour with Schiff's reagent. (The colour is usually formed slowly with propanone.)

78

(b) Phosphorus pentachloride reacts with ketones to produce a *gem* dihalide, but hydrogen chloride is not evolved.

$$\underset{R'}{\overset{R}{\diagdown}}C{=}O + PCl_5 \rightarrow \underset{R'}{\overset{R}{\diagdown}}C\underset{Cl}{\overset{Cl}{\diagup}} + POCl_3$$

*Note:* a *gem* dihalide (*gemini* = twins) is one in which two halogen atoms are carried by the same carbon atom. Vic or vicinal dihalides are those in which the two halogen atoms are located on adjacent carbon atoms.

**Uses of propanone**   Propanone is widely employed as a solvent and as an intermediate in the manufacture of other organic compounds, such as ethanoic anhydride, trichloromethane, etc.

TABLE 12 *Comparison of methanal, ethanal and propanone*

|  | Methanal | Ethanal | Propanone |
|---|---|---|---|
| *Preparation* | oxidation of a primary alcohol | oxidation of a primary alcohol | oxidation of a secondary alcohol |
| *Further oxidation* | readily oxidized to $HCO_2H$ | readily oxidized to $CH_3CO_2H$ | oxidized under drastic conditions to $CO_2$ and $CH_3CO_2H$ |
| *Reducing action* | reduces Fehling's solution and ammoniacal silver nitrate solution | reduces Fehling's solution and ammoniacal silver nitrate solution | none |
| *Action of NaOH* | Cannizzaro's reaction giving $CH_3OH$ and $HCO_2H$ | on heating gives a brown colour due to resin formation | no action |
| *Reaction with sodium hydrogen sulphite solution* | crystalline addition product | crystalline addition product | crystalline addition product |
| *Ammonia* | white solid (hexamine) formed | aldehyde ammonia (an addition product) formed | complex products result |
| *Mechanism of addition* | nucleophilic attack | nucleophilic attack | nucleophilic attack |
| *Polymers* | trioxymethylene | paraldehyde and metaldehyde | does not polymerize readily |
| *Reduction with lithium aluminium hydride* | primary alcohol | primary alcohol | secondary alcohol |
| *Reduction by Clemmensen's method* | alkane | alkane | alkane |

# 8    Monobasic Carboxylic Acids

The general formula of the monobasic carboxylic acids is $R . C \overset{\displaystyle O}{\underset{\displaystyle OH}{\diagup\!\!\!\diagdown}}$

where R represents an alkyl group. The functional group is $-C \overset{\displaystyle O}{\underset{\displaystyle OH}{\diagup\!\!\!\diagdown}}$ which is a combination of a *carb*onyl and a hydr*oxyl*

group, hence the name carboxyl group.

TABLE 13 *Lower monobasic carboxylic acids*

| Name | Formula | M.P. °C | B.P. °C |
|------|---------|---------|---------|
| methanoic acid (formic acid) | $HCO_2H$ | 8·4 | 100·5 |
| ethanoic acid (acetic acid) | $CH_3CO_2H$ | 16·6 | 118·1 |
| propanoic acid | $CH_3CH_2CO_2H$ | −20·3 | 141·2 |
| butanoic acid | $CH_3CH_2CH_2CO_2H$ | (−5) | 162·3 |

**Occurrence**   Methanoic acid is present in the stings of some plants and animals, while ethanoic acid is formed during the 'souring' of wines and beers through the action of micro-organisms. Other acids in this series are also present in plant cells and animal tissues.

## General Methods of Preparation

1 By the oxidation of aldehydes or alcohols:

$$R.CH_2OH \xrightarrow{\ O\ } R.CHO \xrightarrow{\ \ } R.CO_2H$$

Both potassium permanganate and potassium dichromate in dilute sulphuric acid are suitable oxidizing agents.

2 Alkaline hydrolysis of an ester or an amide produces the salt of the acid:

$$R.COOR' + NaOH \rightarrow R.CO_2Na + R'OH$$

$$R.CONH_2 + NaOH \rightarrow R.CO_2Na + NH_3$$

On refluxing either an amide or an ester with sodium hydroxide solution, the sodium salt of the acid is produced. The free acid is liberated by the addition of a mineral acid, such as sulphuric or hydrochloric, and the carboxylic acid may be recovered by distillation or ether extraction. Acid hydrolysis gives the free acid directly.

$$R.COOR' + H_2O \underset{H^+}{\rightleftharpoons} R.CO_2H + R'OH$$

3 By the hydrolysis of a nitrile. The reaction occurs on boiling with either dilute acid or alkali. Several stages are involved. First,

$$R.CN + H_2O \rightarrow R.CONH_2$$

Then, if acid is used,

$$R.CONH_2 + H_2O + HCl \rightarrow R.CO_2H + NH_4Cl$$

or, in the case of alkaline hydrolysis,

$$R.CONH_2 + NaOH \rightarrow R.CO_2Na + NH_3$$

## Special Methods of Preparation

**Methanoic acid**  Sodium methanoate is made by the reaction between superheated sodium hydroxide solution and carbon monoxide at high pressure:

$$CO + NaOH \rightarrow HCO_2Na$$

**Ethanoic acid**  It is manufactured by the catalytic oxidation of ethanal and by the oxidation of pentanes and hexanes derived from petroleum.

## Properties and Reactions of the Monobasic Carboxylic Acids

These acids are colourless liquids. The lower members have pungent, often objectionable odours, and are miscible with water. The miscibility decreases as the relative molecular mass increases. The solution in water is acid to litmus and they have a sour taste.

### Acidity of the Carboxylic Acids

Acids are substances capable of furnishing protons (see Part 3). In organic compounds, acidity is found among molecules having a hydroxyl group. The strength of the organic acid depends on:

(i) the ease of ionization of the O—H link
(ii) the stability of the resulting anion.

81

The oxygen atom present in the carbonyl group tends to pull electrons towards itself thus weakening both the O—H and the C—H bonds as shown.

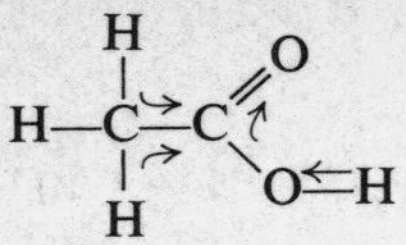

This enhances the ionization of the O—H bond

The carboxylate anion formed on ionization has a special degree of stability as the electrons in the $-CO_2^-$ group can be accommodated in a delocalized orbital system (see Part 1) represented below.

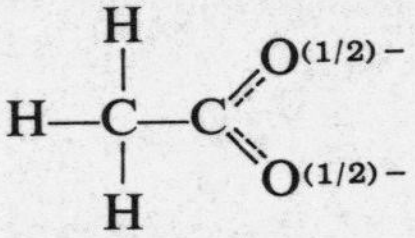

Therefore, carboxylic acids are weak acids. $K_a$ (see Part 3) for methanoic acid is $2.4 \times 10^{-4}$ mol dm$^{-3}$ (or p$K_a$, which equals $-\log K_a$, is 3.62). Similarly, $K_a$ for ethanoic acid is $1.8 \times 10^{-5}$ mol dm$^{-3}$ or p$K_a$ is 4.74. It is interesting to note that the p$K_a$ values for the chlorinated ethanoic acids are:

monochloroethanoic acid p$K_a$ = 2.85
dichloroethanoic acid p$K_a$ = 1.29
trichloroethanoic acid p$K_a$ = 0.65

(A lower p$K_a$ value indicates greater acid strength.) This is due to the fact that the chlorine is a highly electronegative element and the O—H link is weakened by the pull on the hydroxyl electrons exerted by the chlorine atoms (see Figure 10).

The chemical properties are chiefly concerned with the reactions of the carboxyl group. Although this group contains a carbonyl group, reactions associated with that group such as addition and condensation are not observed. Substitution in the alkyl group is again evident. The reactions of ethanoic acid are given as typical of the series.

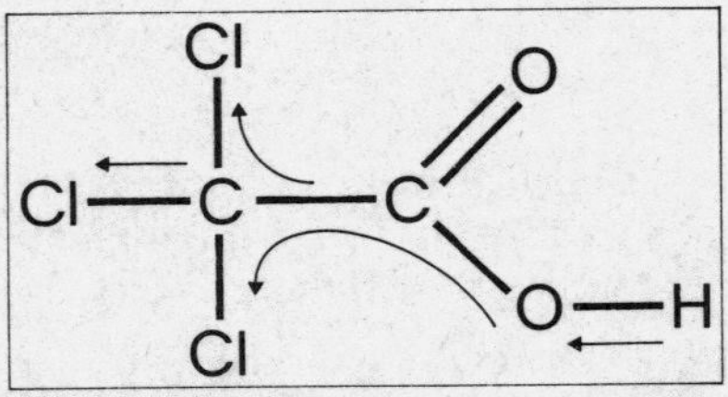

Fig. 10 Enhancement of the acidic properties of trichloroethanoic acid

1 Action as acids. The acid strength gradually decreases among the homologues as the relative molecular mass increases. These acids affect indicators and

(a) neutralize alkalies, forming salts:

$$CH_3CO_2H + NaOH \rightarrow CH_3CO_2Na + H_2O$$

(b) liberate carbon dioxide from carbonates:

$$2CH_3CO_2H + Na_2CO_3 \rightarrow CO_2 + H_2O + 2CH_3CO_2Na$$

(c) evolve hydrogen during reaction with the very reactive metals such as magnesium:

$$2CH_3CO_2H + Mg \rightarrow (CH_3CO_2)_2Mg + H_2$$

2 The hydroxyl part of the carboxyl group may be substituted by chlorine, or eliminated as water on esterification.

(a) Phosphorus pentachloride (or trichloride) reacts vigorously with ethanoic acid to produce ethanoyl chloride:

$$CH_3C\!\!\begin{array}{c} \nearrow O \\ \searrow OH \end{array} + PCl_5 \rightarrow POCl_3 + HCl + CH_3C\!\!\begin{array}{c} \nearrow O \\ \searrow Cl \end{array}$$

Methanoic acid reacts differently:

$$HCO_2H + PCl_5 \rightarrow 2HCl + POCl_3 + CO$$

(b) On refluxing an acid with an alcohol, in the presence of a little concentrated sulphuric acid, a condensation type of reaction takes place, leading to the formation of an ester:

$$CH_3C\!\!\begin{array}{c} \nearrow O \\ \searrow \underline{OH} \end{array} + \begin{array}{c} R \\ | \\ \underline{OH} \end{array} \rightleftharpoons H_2O + CH_3C\!\!\begin{array}{c} \nearrow O \\ \searrow OR \end{array}$$

3 Boiling ethanoic acid reacts with chlorine especially in bright sunlight to give a succession of substitution products:

$$CH_3CO_2H + Cl_2 \rightarrow HCl + CH_2Cl.CO_2H$$
monochloroethanoic acid

$$CH_2ClCO_2H + Cl_2 \rightarrow HCl + CHCl_2.CO_2H$$
dichloroethanoic acid

$$CHCl_2CO_2H + Cl_2 \rightarrow HCl + CCl_3.CO_2H$$
trichloroethanoic acid

Methanoic acid is not substituted in this way.

**Special properties of methanoic acid**  This acid contains an aldehyde grouping.

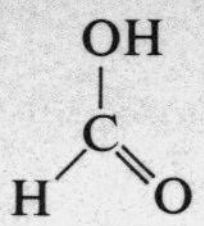

$$\underset{\underset{H}{\overset{\displaystyle |}{}} \overset{\displaystyle OH}{\underset{\displaystyle }{C}} }{\phantom{x}}\,O$$

and some of the reactions of aldehydes are given by methanoic acid. In particular, the acid has some reducing powers, as it decolourizes potassium permanganate solution, gives the silver mirror test and reduces mercury (II) ion to mercury (I). However, it does not reduce Fehling's solution.

Concentrated sulphuric acid dehydrates methanoic acid to give carbon monoxide:

$$HCO_2H - H_2O \rightarrow CO$$

## Summary of the Reactions of Ethanoates and Methanoates

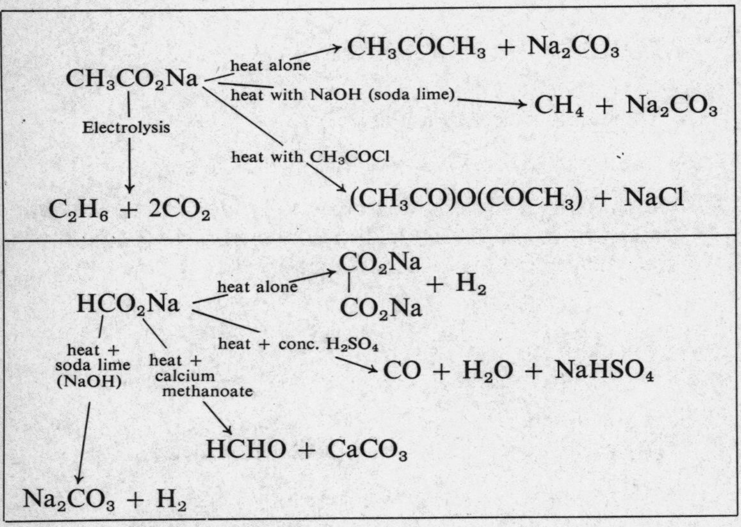

$CH_3CO_2Na$

- heat alone $\rightarrow CH_3COCH_3 + Na_2CO_3$
- heat with NaOH (soda lime) $\rightarrow CH_4 + Na_2CO_3$
- Electrolysis $\downarrow$ $C_2H_6 + 2CO_2$
- heat with $CH_3COCl$ $\rightarrow (CH_3CO)O(COCH_3) + NaCl$

$HCO_2Na$

- heat alone $\rightarrow \begin{matrix} CO_2Na \\ | \\ CO_2Na \end{matrix} + H_2$
- heat + conc. $H_2SO_4$ $\rightarrow CO + H_2O + NaHSO_4$
- heat + calcium methanoate $\rightarrow HCHO + CaCO_3$
- heat + soda lime (NaOH) $\downarrow$ $Na_2CO_3 + H_2$

# 9  Acid Derivatives

Four types of acid derivative are recognized—acid chlorides, amides, anhydrides and esters. They are called acid derivatives since they are produced by reactions in which a complete replacement of the hydroxyl group in an acid takes place.

## ACYL CHLORIDES (ACID CHLORIDES)

In these compounds, the hydroxyl group has been replaced by chlorine, leading to the general formula

$$R.C{\overset{\displaystyle O}{\underset{\displaystyle Cl}{<}}}$$

The R.CO— group is known as an *acyl* group.

Methanoyl chloride does not exist at room temperature, and ethanoyl (acetyl) chloride, $CH_3COCl$, the first member of the series, is quoted in the examples below as showing properties typical of the series.

### General Methods of Preparation

Acyl chlorides are produced by the action of phosphorus trichloride (or pentachloride) on the corresponding acid:

$$3CH_3CO_2H + PCl_3 \rightarrow 3CH_3COCl + H_3PO_3$$

Alternatively, the use of thionyl chloride may give products which are more easily separated:

$$R.CO_2H + SOCl_2 \rightarrow SO_2 + HCl + R.COCl$$

Acyl chlorides decompose in the presence of water, so that the apparatus in which the preparative reactions are effected must be completely dry. The products are separated by distillation.

### Properties and Reactions of Ethanoyl Chloride

Ethanoyl chloride is a colourless liquid which fumes in moist air (owing to hydrolysis). It has a very irritating odour.

Ethanoyl (acetyl) chloride is very reactive and many of the reactions involve the elimination of the chlorine atom, usually as hydrogen chloride. In this way, the $CH_3CO-$ group (the ethanoyl group) is introduced into the other reacting molecule, and the reactions in which this is accomplished are termed ethanoylation (or acetylation, or, in general terms, acylation) reactions.

1 Ethanoyl chloride is vigorously hydrolysed by water:

$$CH_3COCl + H_2O \rightarrow CH_3CO_2H + HCl$$

2 It ethanoylates alcohols forming esters:

$$CH_3COCl + R.OH \rightarrow CH_3COOR + HCl$$

3 It reacts with ammonia, producing ethanamide:

$$CH_3COCl + NH_3 \rightarrow HCl + CH_3CONH_2$$
$$\text{ethanamide}$$

4 Phenylamine is converted into ethanoyl phenylamine:

$$CH_3COCl + C_6H_5NH_2 \rightarrow CH_3CONHC_6H_5 + HCl$$

5 On distilling with sodium ethanoate, ethanoic anhydride is formed:

$$CH_3COCl + CH_3CO_2Na \rightarrow NaCl + (CH_3CO)-O-(COCH_3)$$

6 Ethanoyl chloride is reduced by hydrogen in the presence of a partially poisoned palladium catalyst (p. 65) to give ethanal:

$$CH_3COCl + H_2 \rightarrow HCl + CH_3CHO$$

7 Since ethanoyl chloride is readily decomposed by water, addition of a drop of ethanoyl chloride to silver nitrate solution (in dilute nitric acid) produces an immediate precipitate of silver chloride. Acyl bromides and iodides may also be prepared from acids by the action of the appropriate phosphorus halide.


## ACID ANHYDRIDES

These are produced by the elimination of a molecule of water from two molecules of acid. The general formula is

$$R-\underset{\underset{O}{\|}}{C}-O-\underset{\underset{O}{\|}}{C}-R$$

Ethanoic anhydride (where R represents a methyl group) is the simplest member of the series.

**Preparation**

By the action of ethanoyl chloride on the sodium salt of an acid:

$$CH_3COCl + CH_3CO_2Na \rightarrow NaCl + \begin{array}{c} CH_3C \diagup^{\textstyle O} \diagdown O \\ \diagup \\ CH_3C \diagdown_{\textstyle O} \end{array}$$

By this method simple anhydrides (containing identical alkyl groups) and mixed anhydrides (different alkyl groups) may be prepared.

Ethanoic anhydride is manufactured from ethanoic acid (or from propanone) as it is used in the production of cellulose acetate. The methods used involve the production of a reactive intermediate called ketene.

$$CH_3CO_2H \xrightarrow[\text{with a catalyst}]{\text{heat at reduced pressure}} CH_2{=}C{=}O + H_2O$$
$$\text{ketene}$$

or,

$$CH_3COCH_3 \xrightarrow[750° C]{\text{heat to}} CH_2{=}C{=}O + CH_4$$

Then,

$$CH_2{=}C{=}O + CH_3CO_2H \rightarrow \underset{\displaystyle O}{CH_3C}{-}O{-}\underset{\displaystyle O}{CCH_3}$$

**Properties and Reactions of Ethanoic Anhydride**

Ethanoic anhydride is a colourless liquid which has an irritating smell. It is not readily miscible with water. It is a very reactive substance, and is used as an acetylating agent.

1 It is hydrolysed by water. In view of its limited miscibility, hydrolysis is sometimes a delayed action, although once started, it can be vigorous.

$$\begin{array}{c} CH_3CO \\ \diagdown \\ \diagup \\ CH_3CO \end{array} O + H_2O \rightarrow 2CH_3CO_2H$$

87

2 It ethanoylates (acetylates) alcohols forming esters:

$$\begin{array}{c} CH_3CO \\ {\diagdown} \\ \phantom{CH_3}O + R.OH \rightarrow CH_3COOR + CH_3CO_2H \\ {\diagup} \\ CH_3CO \end{array}$$

Cellulose molecules contain many hydroxyl groups and this substance is ethanoylated by means of a mixture of ethanoic anhydride and ethanoic acid to form cellulose acetate, a substance widely used in the production of man-made fibres.

3 Ethanoic anhydride reacts with ammonia to form ethanamide:

$$\begin{array}{c} CH_3CO \\ {\diagdown} \\ \phantom{CH_3}O + NH_3 \rightarrow CH_3CO_2H + CH_3CONH_2 \\ {\diagup} \\ CH_3CO \end{array}$$

4 Phenylamine is converted into ethanoyl phenylamine:

$$\begin{array}{c} CH_3CO \\ {\diagdown} \\ \phantom{CH_3}O + C_6H_5NH_2 \rightarrow CH_3CO_2H + C_6H_5NHCOCH_3 \\ {\diagup} \\ CH_3CO \end{array}$$

5 Ethanoic anhydride is reduced by lithium aluminium hydride to give ethanol:

$$\begin{array}{c} CH_3C{-}O{-}CCH_3 + 6(H) \rightarrow 2CH_3CH_2OH \\ {\parallel} \quad\quad {\parallel} \\ O \quad\quad\; O \end{array}$$

Ethanoic anhydride may be distinguished from ethanoyl chloride as the products of hydrolysis do not form a precipitate on the addition of silver nitrate solution and dilute nitric acid.

## AMIDES

The general formula of the amides is

$$\begin{array}{c} \phantom{R-C}O \\ \phantom{R-C}{\diagup\!\!\!=} \\ R{-}C \\ \phantom{R-C}{\diagdown} \\ \phantom{R-C}NH_2 \end{array}$$

the hydroxyl group of the acid being replaced by the —NH$_2$ group. R represents hydrogen or an alkyl radical, so that the simplest member of the series is methanamide, HCONH$_2$. In the following notes, the preparation and properties of ethanamide will be discussed as characteristic of the series.

88

## Preparation

1 By the slow dehydration of ammonium salts of carboxylic acids, on gentle heating in the presence of ethanoic acid:

$$CH_3CO_2NH_4 \rightarrow CH_3CONH_2 + H_2O$$

2 Ethanoylation of ammonia:

$$CH_2COCl + NH_3 \rightarrow HCl + CH_3CONH_2$$

$$(CH_3CO)_2O + NH_3 \rightarrow CH_3CO_2H + CH_3CONH_2$$

3 From esters, on treatment with a solution of ammonia in water or alcohol at room temperature:

$$CH_3COOR + NH_3 \rightarrow CH_3CONH_2 + ROH$$

This reaction is known as *ammonolysis*.

## Properties and Reactions of Ethanamide

Ethanamide is a white crystalline solid (methanamide is a liquid) which is soluble in water. It has a 'musty' odour.

1 Ethanamide is hydrolysed on boiling with either acid or alkali:

$$CH_3CONH_2 + HCl + H_2O \rightarrow CH_3CO_2H + NH_4Cl$$

$$CH_3CONH_2 + NaOH \qquad \rightarrow CH_3CO_2Na + NH_3$$

2 It is dehydrated by phosphorus pentoxide to form a nitrile:

$$CH_3CONH_2 - H_2O \rightarrow CH_3CN$$

3 On treatment with ice-cold nitrous acid (i.e. a solution of sodium nitrite in dilute hydrochloric acid) nitrogen is evolved:

$$CH_3CONH_2 + HNO_2 \rightarrow N_2 + CH_3CO_2H + H_2O$$

The evolution of nitrogen with nitrous acid is characteristic of the $-NH_2$ group.

4 It is reduced by heating with sodium in alcohol, or with lithium aluminium hydride, or by hydrogen under pressure:

$$CH_3CONH_2 + 2H_2 \rightarrow H_2O + CH_3CH_2NH_2$$
$$\text{ethylamine}$$

5 It reacts with bromine water and warm alkali to form a primary amine. The reaction takes place in two stages:

$$CH_3CONH_2 + Br_2 \rightarrow HBr + CH_3CONHBr$$
$$\text{bromo-ethanamide}$$

Then,

$$CH_3CONHBr + 3NaOH \rightarrow NaBr + Na_2CO_3 + H_2O + CH_3NH_2$$

This reaction is known as *Hofmann's degradation.* (A degradation is a reaction which brings about a decrease in the number of carbon atoms in the organic molecule.)

## ESTERS

Esters have the general formula

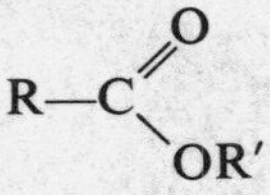

where R may be hydrogen or any alkyl radical, and R' represents an alkyl radical which may or may not be the same as R.

TABLE 14 *Esters*

| Name | Formula | B.P.°C |
|---|---|---|
| methyl methanoate | $HCO.OCH_3$ | 31·7 |
| ethyl methanoate | $HCO.OCH_2CH_3$ | 54·2 |
| methyl ethanoate | $CH_3CO.OCH_3$ | 57·3 |
| ethyl ethanoate | $CH_3CO.OCH_2CH_3$ | 77.2 |

## Preparation

1 By refluxing an alcohol with anhydrous acid, in the presence of concentrated sulphuric or hydrochloric acids

$$R.CO_2H + R'.OH \rightleftharpoons R.COOR' + H_2O$$

This reaction, known as *esterification,* is reversible, but a better yield may be obtained by using an excess of one of the reactants (usually the cheaper one). This method is also used for the production of esters on an industrial scale.

2 Reaction of an alcohol with an acyl chloride or anhydride:

$$R'.OH + R.COCl \rightarrow R.COOR' + HCl$$

3 Silver salt method. The silver salt of the acid is heated with a halogeno-alkane:

$$R.CO_2Ag + R'I \rightarrow R.COOR' + AgI$$

This method often proves lengthy and expensive since it depends on the production of the dry silver salt of the acid.

## Properties and Reactions of Esters

The esters are colourless liquids, which are slightly soluble in water. They have pleasant fruity odours.

Esters are less reactive than other acid derivatives.

1 They are hydrolysed on refluxing with dilute alkalies (or acids).

$$CH_3COOC_2H_5 + NaOH \rightarrow CH_3CO_2Na + C_2H_5OH$$

Hydrolysis is the reverse of esterification. This reaction is also known as *saponification* as the hydrolysis of natural oils and fats (which are esters of glycerol) by alkali results in the formation of soap.

2 Esters form amides with ammonia:

$$R.COOR' + NH_3 \rightarrow R.CONH_2 + R.OH$$

3 They are reduced to alcohols by refluxing with sodium in alcohol (the Bouveault-Blanc method):

$$CH_3COOC_2H_5 + 4(H) \rightarrow 2CH_3CH_2OH$$

Reduction to alcohols may be also effected by high-pressure hydrogenation in the presence of a suitable catalyst.

4 Ester interchange (or alcoholysis). On refluxing an ester with an alcohol and a small amount of the sodium derivative $(RO.Na)$ of the alcohol used, an interchange of alkyl groups occurs. The reaction is reversible and an equilibrium is set up:

$$CH_3COOC_2H_5 + R.OH \rightleftharpoons CH_3COOR + C_2H_5OH$$

The reaction is of importance in the manufacture of complex esters, used in the paint and plastics industries.

**Uses**   Esters are used as flavourings, perfumes, solvents and plasticizers.

## Isomers

In addition to the isomers resulting from differences in the alkyl groups in the ester, every ester is isomeric with at least one acid. For example, isomers of the molecular formula $C_3H_6O_2$ include:

(a) $CH_3COOCH_3$  ⎫
(b) $HCOOC_2H_5$  ⎬ esters
(c) $CH_3CH_2CO_2H$    acid

The acid isomer (c) is easily recognized by its behaviour towards sodium carbonate solution. The two isomeric esters are distinguished by characterizing the products of hydrolysis. In this example, (a) would produce methanol and sodium ethanoate on alkaline hydrolysis, but (b) would give ethanol (which gives the tri-iodomethane or iodoform test) and sodium methanoate (which has reducing properties).

## Mechanism of Esterification

Using methanol containing the heavy isotope of oxygen ($O^{18}$— or $O^*$ in the equation below) it has been shown that esterification proceeds as follows:

$$R.CO_2H + CH_3O^*H \rightarrow R.COO^*CH_3 + H_2O$$

since no heavy oxygen is found in the water produced during the reaction. This indicates that the acid loses its hydroxyl group during esterification.

# 10 Amines, Nitriles and Iso-Nitriles

## AMINES

The amines may be regarded as derivatives of ammonia, the hydrogen atoms being successively replaced by alkyl groups. Three classes of amines exist:

(a) Primary amines    R—N—H    e.g. methylamine    $CH_3$—N—H
                                  |                                        |
                                  H                                        H

(b) Secondary amines    R—N—R    e.g. dimethylamine    $CH_3$—N—$CH_3$
                                      |                                        |
                                  H                                        H

(c) Tertiary amines    R—N—R    e.g. trimethylamine    $CH_3$—N—$CH_3$
                                      |                                        |
                                  R                                        $CH_3$

*Note:* the alkyl groups in these compounds may or may not be the same.

In addition, ionic, solid, quaternary ammonium salts are formed; for example, $(CH_3)_4N^+I^-$, tetramethylammonium iodide.

The functional group of a primary amine is the —$NH_2$ group.

### Preparation of Primary Amines

1 By Hofmann's degradation of an amide:

$$R.CONH_2 + Br_2 + 4NaOH \rightarrow 2NaBr + Na_2CO_3$$
$$+ 2H_2O + R.NH_2$$

This method yields primary amines only.

2 By the reduction of a nitrile using sodium and alcohol:

$$R.CN + 2H_2 \rightarrow R.CH_2NH_2$$

3 By the reduction of an amide using lithium aluminium hydride or sodium and alcohol:

$$R.CONH_2 + 2H_2 \rightarrow R.CH_2NH_2 + H_2O$$

## Preparation of Secondary and Tertiary Amines

Secondary amines are produced on the reduction of an iso-nitrile:

$$R.NC + 2H_2 \rightarrow R.NHCH_3$$

The reaction of ammonia with a halogeno-alkane in the correct proportions also gives rise to secondary and tertiary amines:

$$2R.X + NH_3 \rightarrow R_2N^+H_2X^- + HX$$

$$3R.X + NH_3 \rightarrow R_3N^+HX^- + 2HX$$

Dimethylamine, required in the production of solvents for synthetic fibres, is made on an industrial scale by the high-pressure reaction between methanol and ammonia at 400°C in the presence of an alumina catalyst:

$$2CH_3OH + NH_3 \rightarrow (CH_3)_2NH + H_2O$$

The reaction products contain methylamine and trimethylamine in addition to dimethylamine.

Quaternary ammonium salts, used in the manufacture of detergents and disinfectants are produced by heating ammonia with a large excess of a halogeno-alkane.

$$4R.X + NH_3 \rightarrow 3HX + R_4N^+X^-$$

## Properties and Reactions of Amines

Methylamine, di- and tri-methylamine are gases. Ethylamine (b.p. 16·8°C) and the other lower amines are either gases or very volatile liquids. They resemble ammonia in being basic and very soluble in water to yield alkaline solutions. They have a strong odour, smelling of fish and ammonia.

Chemical properties are as follows:

1 They burn in air, unlike ammonia:

$$4CH_3NH_2 + 9O_2 \rightarrow 4CO_2 + 10H_2O + 2N_2$$

2 A solution in water is alkaline:

$$CH_3NH_2 + H_2O \rightleftharpoons CH_3NH_3^+OH^-$$

(Compare $NH_3 + H_2O \rightleftharpoons NH_4^+OH^-$.)

3 They form salts with acids (analogous to ammonium salts):

$$CH_3NH_2 + HCl \rightarrow CH_3NH_3^+Cl^-$$

methylammonium chloride

These salts are decomposed by sodium hydroxide, liberating the free base:

$$CH_3NH_3{}^+Cl^- + NaOH \rightarrow CH_3NH_2 + Na^+Cl^- + H_2O$$

The ionic nature of chlorine in these salts is shown by the formation of a white precipitate when silver nitrate solution is added.

4 A solution of an amine in water precipitates metal hydroxides, for example:

$$2CH_3NH_3{}^+OH^- + ZnSO_4 \rightarrow (CH_3NH_3{}^+)_2SO_4{}^{2-} + Zn(OH)_2$$

5 Primary and secondary amines react with acyl chlorides or anhydrides to give substituted amides:

$$R.COCl + CH_3NH_2 \rightarrow R.CONHCH_3 + HCl$$

$$R.COCl + (CH_3)_2NH \rightarrow R.CON(CH_3)_2 + HCl$$

6 When primary amines are heated with trichloromethane and alkali, the foul-smelling iso-nitrile is produced:

$$CHCl_3 + 3KOH + CH_3NH_2 \rightarrow 3KCl + 3H_2O + CH_3NC$$

methyl isonitrile

Only the primary amines give this test.

7 The reaction with ice-cold nitrous acid is used to distinguish between primary, secondary and tertiary amines. An ice-cold solution of sodium nitrite is acidified with dilute hydrochloric acid to give a solution of nitrous acid *in situ*.

Primary amines liberate nitrogen gas in quantitative yield.

$$CH_3CH_2NH_2 + HNO_2 \rightarrow CH_3CH_2OH + H_2O + N_2$$

(N.B. Methylamine does not produce methanol.)

With secondary amines, a yellow oil (a nitrosamine) is formed, and no nitrogen is produced:

$$(CH_3)_2NH + HNO_2 \rightarrow (CH_3)_2NNO + H_2O$$

Tertiary amines show no reaction other than dissolving in the acid, and again, no nitrogen is evolved.

**Isomerism**  Amines show metamerism, for example,

$$\underset{\underset{H}{|}}{C_2H_5-N-C_2H_5} \quad \text{and} \quad \underset{\underset{H}{|}}{CH_3-N-C_3H_7}$$

# NITRILES AND ISO-NITRILES

The nitriles have the general formula RCN, while that of the iso-nitriles is RNC. The methods of preparation and chief properties are summarized in Table 15.

TABLE 15

| | Nitriles | Iso-nitriles |
|---|---|---|
| Formula | $CH_3CN$ | $CH_3NC$ |
| Name | cyano-methane or ethanonitrile | ethanoisonitrile |
| Preparation | 1. by refluxing iodomethane with potassium cyanide in aqueous alcohol: $CH_3I + KCN \rightarrow KI + CH_3CN$ | 1. by refluxing iodomethane with silver cyanide in aqueous alcohol: $CH_3I + AgCN \rightarrow AgI + CH_3NC$ |
| | 2. by dehydration of an amide using phosphorus pentoxide: $CH_3CONH_2—H_2O \rightarrow CH_3CN$ | 2. from a primary amine by reacting with trichloromethane and alkali: $CHCl_3 + 3KOH + CH_3NH_2 \rightarrow 3KCl + 3H_2O + CH_3NC$ |
| Properties | colourless, sweet-smelling soluble in water (higher members insoluble) and slightly poisonous | colourless liquid with a highly objectionable odour; very poisonous |
| | hydrolysed by boiling with sodium hydroxide solution, or moderately concentrated sulphuric acid $CH_3CN + 2H_2O \rightarrow CH_3CO_2H + NH_3$ | hydrolysed slowly by acids: $CH_3NC + 2H_2O + HCl \rightarrow CH_3NH_3{}^+Cl^- + HCO_2H$ |
| | reduced by lithium aluminium hydride, or sodium in alcohol to give a primary amine: $CH_3CN + 2H_2 \rightarrow CH_3CH_2NH_2$ | reduction under similar conditions gives a secondary amine: $CH_3NC + 2H_2 \rightarrow CH_3NHCH_3$ |

# 11  The Aromatic Hydrocarbons

All the compounds considered so far in this book have been open chain or *aliphatic* compounds. Another important class of compounds is the *aromatic* series, the simplest hydrocarbon in this class is benzene.

## BENZENE

Benzene is present in both crude coal gas and in coal tar. It is extracted from the former by passing the gas through a spray of oil in which the benzene dissolves. The solution of benzene in oil is fractionally distilled, when benzene and methylbenzene (toluene) (also found in coal gas) are obtained.

Coal tar is first distilled, and the lowest-boiling fraction (known as light oil) is first treated with 10% sodium hydroxide solution and then washed with 25% sulphuric acid. The resulting light oil is fractionated and yields benzene, toluene and several other aromatic compounds of low boiling point.

Aromatic hydrocarbons are also manufactured on a large scale by the catalytic dehydrogenation of the corresponding alkanes found in petroleum. For example,

$$C_6H_{14} \rightarrow C_6H_6 + 4H_2$$

### Preparation of Benzene

1 By the decarboxylation of benzoates on heating with soda lime:

$$C_6H_5CO_2Na + NaOH \rightarrow Na_2CO_3 + C_6H_6$$

2 From benzene diazonium chloride on boiling with alcohol:

$$C_6H_5N_2Cl + C_2H_5OH \rightarrow N_2 + HCl + CH_3CHO + C_6H_6$$

3 By heating benzene sulphonic acid with high-pressure steam:

$$C_6H_5SO_3H + H_2O \rightarrow H_2SO_4 + C_6H_6$$

4 From phenol vapour by reduction with hot zinc powder:

$$C_6H_5OH + Zn \rightarrow C_6H_6 + ZnO$$

## Properties and Reactions of Benzene

Benzene is a colourless liquid having a characteristic rather pleasant odour. It is toxic, and prolonged inhalation of benzene vapour can be the cause of anaemia. It is immiscible with water (forming an upper layer) and solidifies at 5·5°C. Benzene may be purified by crystallization, and this is done on a commercial scale.

Chemical properties and reactions are:

1 Benzene burns in air with a very sooty flame.

In excess air or oxygen:

$$2C_6H_6 + 15O_2 \rightarrow 12CO_2 + 6H_2O$$

In a limited supply of air:

$$2C_6H_6 + 3O_2 \rightarrow 6H_2O + 12C$$

Benzene shows a great resistance to attack by oxidizing agents. Its formula, $C_6H_6$, suggests a high degree of unsaturation, yet the number of addition reactions are few, the chief type of reaction being substitution.

2 *The addition reactions* of benzene are as follows.

(a) Addition of hydrogen takes place at 200°C and over a nickel catalyst:

$$C_6H_6 + 3H_2 \rightarrow C_6H_{12}$$

The product is cyclohexane, which is a colourless liquid having the characteristic properties of an alkane. It is, in fact, a cyclic alkane.

(b) On passing chlorine through boiling benzene in bright sunlight, or in ultraviolet light, benzene hexachloride is formed:

$$C_6H_6 + 3Cl_2 \rightarrow C_6H_6Cl_6$$

One isomer of benzene hexachloride is used as a powerful insecticide. Bromine reacts in a similar manner.

(c) Benzene tri-ozonide is formed when ozonized air is passed through benzene at room temperature:

$$C_6H_6 + 3O_3 \rightarrow C_6H_6(O_3)_3$$

3 *The substitution reactions* are as follows.

(a) Passing chlorine through boiling benzene in subdued light and in the presence of a catalyst such as iron or iodine (known as halogen carriers), a series of substitution products are formed:

$$C_6H_6 + Cl_2 \rightarrow HCl + C_6H_5Cl$$
monochlorobenzene

$$C_6H_5Cl + Cl_2 \rightarrow HCl + C_6H_4Cl_2$$
dichlorobenzene

The process continues until eventually hexachlorobenzene, $C_6Cl_6$, is produced.

(b) On heating with concentrated sulphuric acid under reflux for some time, benzene sulphonic acid is produced:

$$C_6H_6 + H_2SO_4 \rightarrow H_2O + C_6H_5 . SO_3H$$

(c) On shaking with a mixture of concentrated nitric and sulphuric acids, keeping the temperature below 60°C, nitrobenzene is formed:

$$C_6H_6 + HNO_3 \rightarrow C_6H_5NO_2 + H_2O$$

When fuming nitric acid is used, and the temperature is allowed to rise above 60°C, di- and tri-nitrobenzene are produced successively:

$$C_6H_6 + 2HNO_3 \rightarrow 2H_2O + C_6H_4(NO_2)_2$$

$$\downarrow HNO_3$$

$$C_6H_3(NO_2)_3$$

4 *The Friedel–Crafts reaction:* in the presence of a significant quantity of anhydrous aluminium chloride, benzene reacts with chloroalkanes to give higher aromatic hydrocarbons. For example,

$$C_6H_6 + CH_3Cl \rightarrow HCl + C_6H_5CH_3$$
methylbenzene
(toluene)

Normally, the halogen compound used in this reaction is aliphatic, while the hydrocarbon is aromatic. Unfortunately, a mixture of products is formed corresponding to further substitution (e.g. dimethylbenzene and trimethylbenzene in the above reaction). Catalysts other than aluminium chloride are used and alternative products may result according to the conditions. Another complicating factor is that the alkyl group of the halogen compound may undergo rearrangement during the reaction (for example both 1-bromopropane and 2-bromopropane react with benzene to give the same product).

## The Structural Formula of Benzene

1 Analysis shows benzene to have the empirical formula CH.
2 The relative molecular mass is 78·11 which shows the molecular formula to be $C_6H_6$.

3 No isomers are found with any mono-substituted derivative of benzene. This shows that benzene cannot be a straight-chain compound.

4 As a result of its addition reactions, benzene adds on a maximum of six hydrogen or six chlorine atoms. Hence, the degree of unsaturation is limited to the equivalent of three double bonds, or two triple bonds.

5 The formation of a triozonide suggests that three double bonds are present.

6 The resistance of benzene to attack by oxidizing agents such as acidified potassium permanganate, and the fact that hydrogen halides do not add on to benzene indicate that the double bonds have a character different from those in the alkenes.

7 X-ray analysis shows that the distances between adjacent carbon atoms are identical (0·139 nm), and intermediate between single and double bond lengths.

A cyclic structure which agrees with all these observations was proposed by Kekulé in 1865:

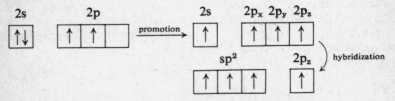

as an alternative molecular structure.

Modern valency theory suggests that three of the four valencies of each carbon atom result from $sp^2$ hybridization (see Part 1, *General and Inorganic Chemistry*).

Carbon has the electronic configuration $1s^2\ 2s^2\ 2p^2$, and by the promotion of an electron from the 2s to the 2p level, four unpaired electrons are available for bonding.

The initial planar structure is formed by the overlapping of the $sp^2$ hybrid orbitals, which also takes account of the carbon–hydrogen bond. The singly filled $p_z$ orbitals of the six carbon atoms merge to

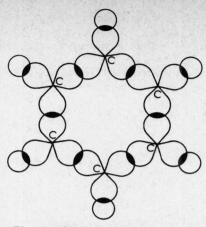

Fig. 11 sp² hybrid orbitals in benzene

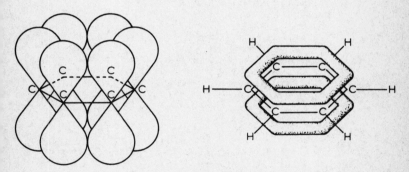

Fig. 12 Formation of a delocalized orbital in benzene

form a large delocalized orbital system lying above and below the benzene nucleus as shown in Figure 12. This interpretation of the bonding in benzene is important in considering the mechanism of the reactions of benzene.

Thus, the symbol used to represent a benzene molecule is

where each corner of the hexagon corresponds to $=\overset{\displaystyle C}{\underset{\displaystyle }{}}\diagdown^{H}$

## Further Substitution Products of Benzene

The introduction of more than one substituent in benzene leads to the formation of isomers. In order to refer to the different isomers without ambiguity, the following two systems are used:

(a) For di-substituted products, the prefixes *ortho-*, *meta*, or *para-* are used according to the relative positions of the two substituents.

(b) Alternatively, the carbon atoms in the benzene nucleus are numbered in such a way as to keep the numbers as small as possible.

Taking the dichlorobenzenes as an example, three isomers are possible, and these are named as shown.

ortho-dichlorobenzene (written o-dichlorobenzene) or
1,2-dichlorobenzene
    (Note that this is the same as 1,6-dichlorobenzene)

meta-dichlorobenzene (written m-dichlorobenzene) or
1,3-dichlorobenzene
    (Note that this is the same as 1,5-dichlorobenzene)

para-dichlorobenzene (written p-dichlorobenzene) or
1,4-dichlorobenzene

Four isomers of trichlorobenzene exist. Each of the above isomers gives rise to a different number of further isomers on tri-substitution.

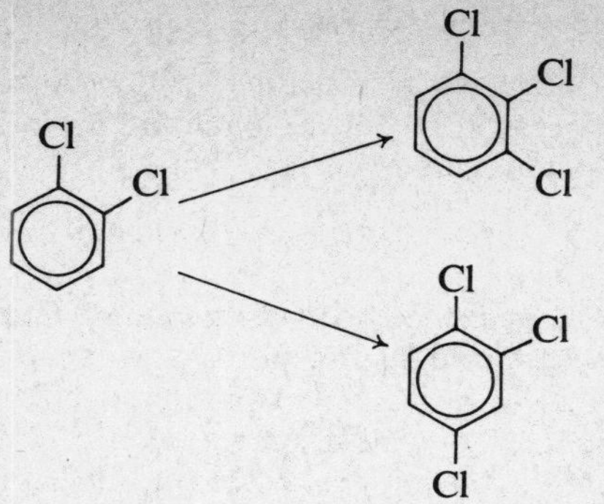

Cl

Cl

1,2,3-trichlorobenzene

Cl

Cl

Cl

1,2,4-trichlorobenzene

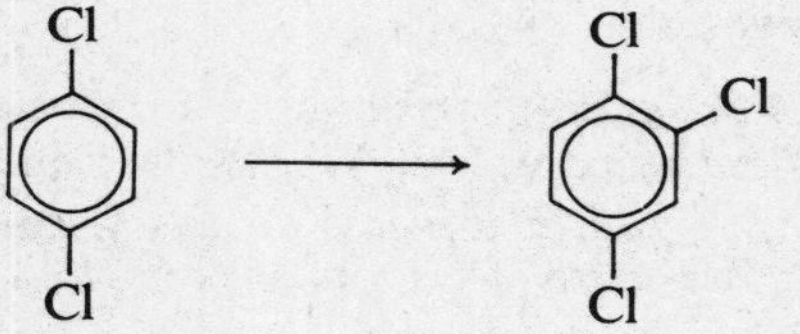

Cl

Cl

Cl

Cl

Cl

1,2,4-trichlorobenzene

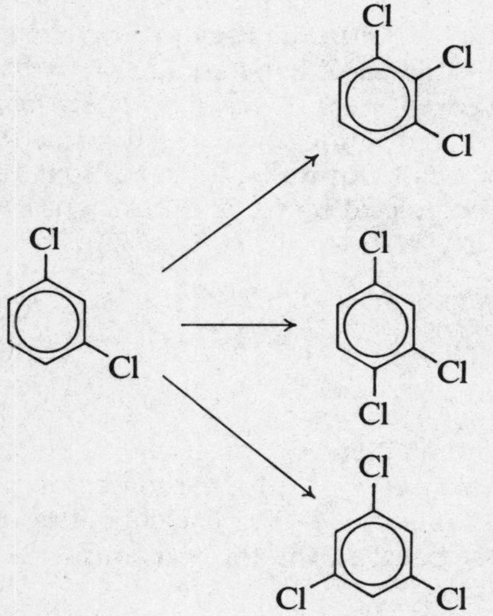

Cl

Cl

Cl

1,2,3-trichlorobenzene

Cl

Cl

Cl

Cl

Cl

1,3,4-trichlorobenzene

Cl

Cl

Cl

Cl

1,3,5-trichlorobenzene

The ortho- isomer produces two tri-substituted isomers, the meta-three and the para- one. If it is practicable to prepare and separate these tri-substituted products, this provides a method of proving the identity of the original di-substituted isomer. This technique is known as Körner's absolute method of isomer identification.

## Outline of the Theory of Substitution Reactions of Benzene

The substitution reactions of benzene (especially nitration) have been extensively studied and three important aspects require an interpretation.

(a) Benzene is attacked by reagents that seek electrons—electrophilic reagents.

(b) Further substitution in benzene is sometimes easier than substitution where there is no initial substituent (in which case the benzene nucleus is said to be activated) or more difficult (in which case the nucleus is said to be deactivated).

(c) According to the nature of the initial substituent, further attack takes place either at the meta position, or at the ortho and para positions. This is known as the *directive effect*.

The presence of a delocalized orbital in benzene which accommodates six electrons (in a system of enhanced stability) means that an excess of electrons over the minimum required for single bonding is present in the benzene molecule. Consequently, electrophiles attack the benzene molecule. Suppose the electrophile is represented as $Z^+$. This is attracted by and then becomes attached to the delocalized orbital. A transition state ensues in which $Z^+$ attaches itself to a particular carbon atom and the strength of the C—Z bond gradually builds up as the C—H bond gradually weakens. During this time, the delocalized orbital system is broken (the activation energy of the reaction includes the energy needed to break down the delocalization) and reforms when substitution is completed by the departure of a proton.

transition state

It can be seen that substitution preserves the delocalized orbital system, whereas a simple addition of $Z^+$ to one carbon atom and the anion (which of necessity accompanies $Z^+$) to a neighbouring carbon atom would destroy it. This explains why the reactions of benzene are predominantly substitution.

**Nitration of benzene** Nitrating acid is a mixture of concentrated nitric and sulphuric acids, and this mixture is known to contain the nitronium ion, $NO_2^+$, produced by the reaction:

$$2H_2SO_4 + HNO_3 \rightleftharpoons NO_2^+ + H_3O^+ + 2HSO_4^-$$

The nitronium ion is an electrophilic reagent.

**Halogenation of benzene**   In the presence of a catalyst, benzene is substituted by halogens. The function of the catalyst is to produce an electrophile from the attacking halogen molecule, for example with bromine Fe (added as a catalyst) gives $FeBr_3$.

Then $\qquad Br\!-\!Br + FeBr_3 \rightarrow \overset{\delta+}{Br}\ldots[Br\ldots FeBr_3]^{\delta-}$

This mechanism is supported by the fact that the interhalogen $\overset{\delta+}{Br}\!-\!\overset{\delta-}{Cl}$ always brominates but never chlorinates as the more electronegative chlorine atom cannot form an electrophilic species in these circumstances. Alkylation by the Friedel–Crafts reaction follows a similar mechanism and it is the function of the aluminium chloride catalyst to produce the attacking electrophile from the halogenoalkane.

$$CH_3I + AlCl_3 \rightarrow \overset{\delta+}{CH_3}\ldots I\ldots \overset{\delta-}{AlCl_3}$$

In sulphonation, the electrophile is probably $SO_3$ produced by the reaction $2H_2SO_4 \rightleftharpoons SO_3 + H_3O^+ + HSO_4^-$

### Activation and Deactivation

The presence of a substituent has a marked effect on the reactivity of benzene derivatives. For example, methylbenzene is more reactive than benzene, but nitrobenzene undergoes further substitution only under drastic conditions. There are two contributory factors underlying this behaviour.

1 *Inductive effect*   The effect of a substituent group is to repel electrons into the benzene ring (called a $+I$ or electron releasing effect) or to attract electrons from the ring ($-I$ effect). An approximate order of the inductive effect is as follows:

$$F \quad OH \quad Cl \quad Br \quad I \quad (H) \quad CH_3 \quad C_2H_5$$
$$\leftarrow(-I)\text{————————— zero ————}(+I)\rightarrow$$

As a result of the $+I$ effect, electrons are released into the benzene ring which facilitates attack by an electrophile. Thus the benzene ring is activated. Deactivation is produced by a substituent group with a $-I$ effect.

2 *Mesomeric effect* When a substituent group contains a double bond or a readily donated lone pair of electrons, it is possible for the delocalized orbital system to extend into the substituent group. This is known as the *mesomeric effect*, and it is this effect which generally governs the relative rates of further substitution *at different positions* in the benzene ring. If the substituent group contains a double bond in such a position that it is separated from the benzene ring by a single bond, the ring is deactivated and further substitution occurs (with difficulty) to give the meta- or 1,3 isomer. However, substituent groups which carry a lone pair on the atom attached directly to the ring, activate the ring and lead to the formation of a mixture of ortho- and para- (1,2 and 1,4) isomers on further substitution. The net results of these effects are summarized in Table 16.

TABLE 16 *Further substitution products of benzene derivatives*

| Deactivated giving 1,3 isomers on substitution | Strongly activated giving a mixture of 1,2 and 1,4 isomers | Weakly activated giving a mixture of 1,2 and 1,4 isomers |
|---|---|---|

An exception is chlorobenzene which gives rise to a mixture of 1,2 and 1,4 isomers on further substitution, but here the strong $-I$ effect withdraws electrons from the ring and deactivates it to a greater extent than the lone pairs on —Cl activate it.

These explanations are a brief outline of the theory of further substitution in benzene. Other factors have to be taken into consideration, and these may modify the general conclusions reached above. Some references to more detailed accounts of this topic are included at the end of this book.

# METHYLBENZENE (TOLUENE)

Methylbenzene has the structural formula

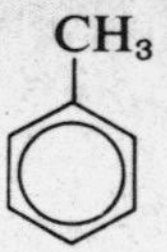

CH₃

and the molecule is made up of a methyl group (referred to as the *side chain*) attached to a benzene nucleus.

## Preparation

1  Methylbenzene is obtained in low yield by heating sodium p-toluate with soda lime:

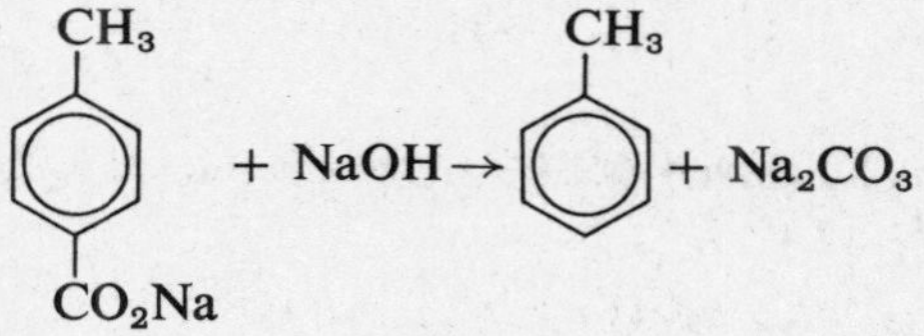

$$\text{p-toluate} + NaOH \rightarrow \text{methylbenzene} + Na_2CO_3$$

2  *Wurtz-Fittig reaction.* Sodium reacts with a mixture of bromo-methane and bromobenzene in dry ethoxy-ethane (ether)

$$C_6H_5Br + CH_3Br + 2Na \rightarrow C_6H_5CH_3 + 2NaBr$$

(Compare with the Wurtz reaction, p. 47.) Side products are ethane and diphenyl ($C_6H_5$—$C_6H_5$).

3  *Friedel–Crafts reaction.* This is an important general method for preparing homologues of benzene. Chloromethane and benzene react by the elimination of hydrogen chloride. The reaction is catalysed by anhydrous aluminium chloride, although this substance has to be present in considerable proportions.

$$C_6H_6 + CH_3Cl \rightarrow HCl + C_6H_5CH_3$$

Methylbenzene is obtained on an industrial scale from coal tar and petroleum sources.

## Properties and Reactions of Methylbenzene

It is a colourless liquid, practically insoluble in water, and having a characteristic aromatic odour. Methylbenzene is considered to be less toxic than benzene and is replacing benzene as a laboratory solvent.

107

Methylbenzene shows two types of reaction, one being concerned with the methyl side chain, in which reactions it behaves in a similar manner to methane. The second type of reaction involves the aromatic nucleus, in which case the reactions are similar to benzene.

1 Methylbenzene burns in air with a very sooty flame. In excess of air or oxygen:

$$C_6H_5CH_3 + 9O_2 \rightarrow 7CO_2 + 4H_2O$$

In a restricted supply of air:

$$C_6H_5CH_3 + 2O_2 \rightarrow 4H_2O + 7C$$

2 Oxidizing agents (for example, refluxing with potassium permanganate solution) convert methylbenzene to benzoic acid.

3 Addition of hydrogen at 200°C over a nickel catalyst produces methyl cyclohexane:

$$C_6H_5CH_3 + 3H_2 \rightarrow C_6H_{11}CH_3$$

4 Substitution reactions in the nucleus.

   (a) Concentrated nitric acid forms a mixture of 2-nitro-methylbenzene and 4-nitro-methylbenzene.

Nitration may be continued using fuming nitric acid or nitrating acid (a mixture of concentrated nitric and sulphuric acids) to yield 2,4-dinitro-methylbenzene and 2,4,6-trinitro-methylbenzene (formerly trinitrotoluene, TNT) respectively:

(b) On refluxing with concentrated sulphuric acid, a mixture of 2- and 4-sulphonic acids is produced:

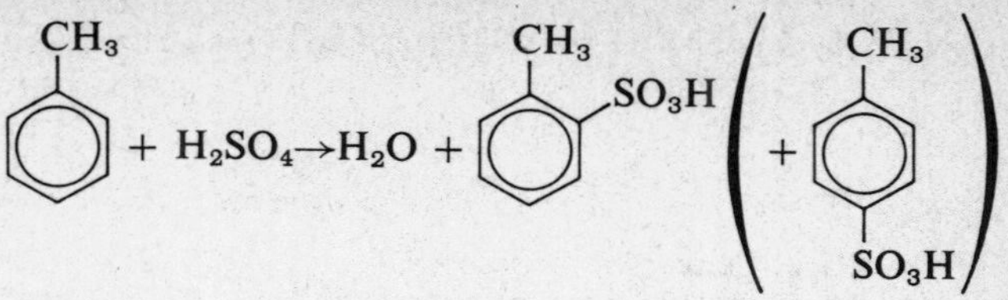

(c) Chlorination of methylbenzene in the cold, using a catalyst such as iron or iodine, gives a mixture of 2- and 4-chloromethylbenzene:

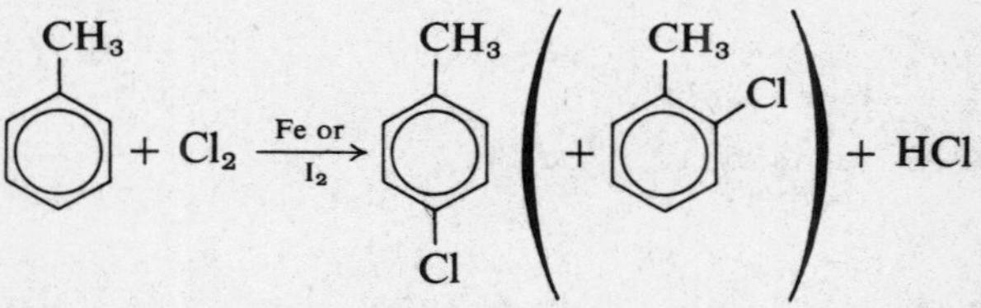

5 Side-chain substitution: on passing chlorine through boiling toluene in bright sunlight or ultraviolet light, a succession of substitution products is formed.

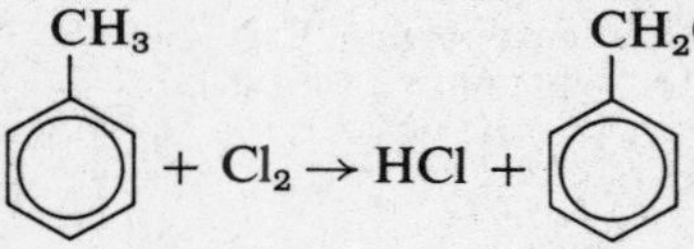

(chloromethyl) benzene
or benzyl chloride

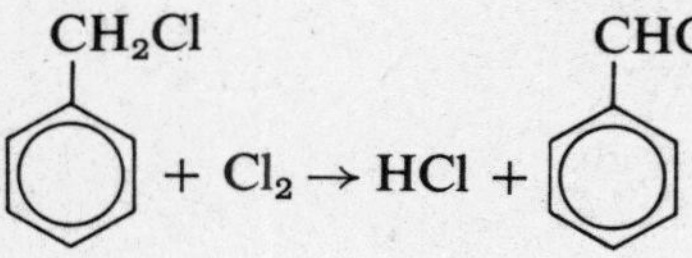

(dichloromethyl) benzene
or benzal chloride

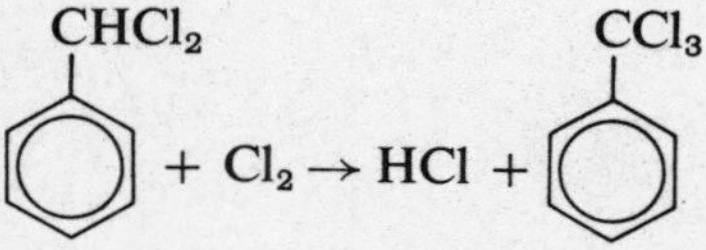

(trichloromethyl) benzene
or benzo-trichloride

These chlorides are very reactive and are readily hydrolysed:

$$C_6H_5CH_2Cl + NaOH \rightarrow NaCl + C_6H_5CH_2OH$$

phenylmethanol
(benzyl alcohol)

$$C_6H_5CHCl_2 + 2NaOH \rightarrow 2NaCl + H_2O + C_6H_5CHO$$

benzaldehyde

$$C_6H_5CCl_3 + 3NaOH \rightarrow 3NaCl + H_2O + C_6H_5CO_2H$$

benzoic acid

TABLE 17 *Comparison of the reactions of halogenated hydrocarbons*

| | *Chloroethane* | *Chlorobenzene* | *(Chloromethyl) benzene* |
|---|---|---|---|
| *Formula* | $C_2H_5Cl$ | $C_6H_5Cl$ | $C_6H_5CH_2Cl$ |
| *Physical properties* | colourless gas; sweetish smell | colourless liquid; faint pleasant odour | colourless liquid; extremely irritating odour |
| *Silver nitrate solution* | precipitate forms slowly | no reaction | precipitate forms rapidly |
| *Dilute alkali* | forms ethanol $C_2H_5OH$ | no reaction | forms phenylmethanol, $C_6H_5CH_2OH$ |
| *Hot concentrated alkali in alcohol* | produces $C_2H_4$ in small yield | forms $C_6H_5OH$ (phenol) when heated to 300°C with NaOH under pressure | forms phenylmethanol, $C_6H_5CH_2OH$ |
| *Ammonia* | with alcoholic ammonia forms $C_2H_5NH_2$ | in presence of a catalyst reacts with aqueous ammonia at 200°C under pressure to give phenylamine, $C_6H_5NH_2$ | produces phenylmethylamine, $C_6H_5CH_2NH_2$, with alcoholic ammonia |

# 12 Aromatic Sulphonic Acids and Phenols

## BENZENE SULPHONIC ACID

The action of concentrated sulphuric acid on aromatic hydrocarbons gives rise to sulphonic acids. Benzene sulphonic acid, produced by refluxing benzene and concentrated sulphuric acid for some time (or by the action of fuming sulphuric acid at room temperature), is a white crystalline solid:

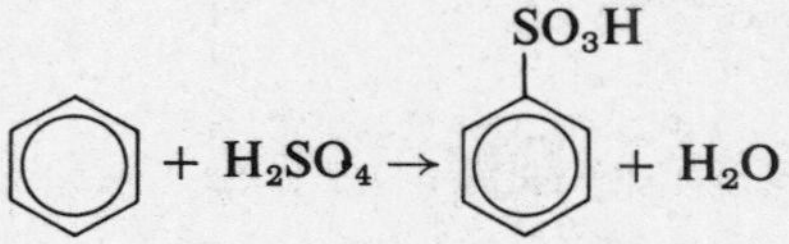

On account of its deliquescent nature, it is usually prepared and stored in the form of its sodium salt, sodium benzene sulphonate. The calcium and barium salts are both soluble.

### Reactions

1 It is a fairly strong acid in solution ($K_a = 0.5$) turning litmus red and liberating carbon dioxide from carbonates. It may be neutralized by alkalies, forming salts, and it also forms esters and amides, etc.

2 When fused with sodium hydroxide, the sodium salt of phenol (sodium phenoxide) is produced:

$$C_6H_5SO_3H + 2NaOH \rightarrow NaHSO_3 + H_2O + C_6H_5ONa$$
$$\text{sodium phenoxide}$$

3 On fusing with potassium cyanide, benzonitrile (phenyl cyanide) is produced:

$$C_6H_5SO_3H + KCN \rightarrow KHSO_3 + C_6H_5CN$$

Benzonitrile (which cannot be made from chlorobenzene) is hydrolysed to ammonium benzoate on boiling with concentrated hydrochloric acid:

$$C_6H_5CN + 2H_2O \rightarrow C_6H_5CO_2NH_4 \xrightarrow{+H^+} C_6H_5CO_2H + NH_4^+$$

111

4 Benzene sulphonyl chloride (which shows the typical reactions of an acid chloride, p. 85) is produced by reaction with phosphorus pentachloride:

$$C_6H_5SO_3H + PCl_5 \rightarrow C_6H_5SO_2Cl + HCl + POCl_3$$

Benzene sulphonyl chloride reacts with ammonia to form benzene sulphonamide:

$$C_6H_5SO_2Cl + NH_3 \rightarrow HCl + C_6H_5SO_2NH_2$$

The sulphonamides and their derivatives form a range of very important drugs.

## PHENOL

Phenols are a class of compounds which have at least one hydroxyl group attached to the aromatic nucleus. Therefore, they may be regarded as the aromatic analogues of the aliphatic alcohols, and some similarities in the reactions of both types of compounds are evident, but the presence of the aromatic nucleus is responsible for some important differences in behaviour. Phenol has the structural formula

### Preparation of Phenol

1 By fusing benzene sulphonic acid with sodium or potassium hydroxide:

$$C_6H_5SO_3H + NaOH \rightarrow NaHSO_3 + C_6H_5OH$$

The fused mass is dissolved in water, and the phenol (present in this reaction mixture as sodium phenoxide) is liberated on acidification. The mixture is then steam distilled and the phenol is extracted from the distillate using ethoxy-ethane (ether).

2 From diazonium salts on boiling with dilute acid:

$$C_6H_5N_2Cl + H_2O \rightarrow N_2 + HCl + C_6H_5OH$$

Phenol is manufactured from benzene by way of benzene sulphonic acid, and from chlorobenzene by hydrolysis at high temperatures and pressures. It is also obtained from coal tar, and a recent method uses petroleum as a source of the chemical from which phenol can be prepared (p. 161).

## Properties and Reactions of Phenol

Phenol is a white crystalline substance which often turns slightly pink on standing. It has a characteristic odour, and is sparingly soluble in cold water, but completely miscible in hot water. It is poisonous and caustic (i.e. burns the skin). Phenols are weakly acidic (p$K_a$ for phenol itself is 10); they scarcely affect pH paper but they do form salts with alkalies (see Table 18).

1 Phenol gives an intense grey-purple colour on mixing with iron (III) chloride solution.
2 The reactions of the aromatic nucleus are as follows:

   (a) It is easily nitrated. Dilute nitric acid produces a mixture of 2- and 4-nitrophenol on reaction with phenol at room temperature:

$$\text{C}_6\text{H}_5\text{OH} + \text{HNO}_3 \rightarrow \text{H}_2\text{O} + \text{(2-nitrophenol)} \quad \left( + \text{(4-nitrophenol)} \right)$$

Concentrated nitric acid, or nitrating acid, forms 2,4,6-trinitrophenol:

$$\text{C}_6\text{H}_5\text{OH} + 3\text{HNO}_3 \rightarrow \text{(2,4,6-trinitrophenol)} + 3\text{H}_2\text{O}$$

This compound, sometimes known as picric acid, is explosive when dry, and is stored under water.

   (b) Sulphonation, using concentrated sulphuric acid, gives 2- and 4-phenolsulphonic acid:

$$\text{C}_6\text{H}_5\text{OH} + \text{H}_2\text{SO}_4 \rightarrow \text{H}_2\text{O} + \text{(4-phenolsulphonic acid, SO}_3\text{H)} \quad \left( + \text{(2-phenolsulphonic acid, SO}_3\text{H)} \right)$$

113

(c) When bromine water is added to phenol, a white precipitate of 2,4,6-tribromophenol is produced:

(d) Phenol may be chlorinated to give 2,4,6-trichlorophenol, but the process may be modified to produce 2,4-dichlorophenol, which is an important intermediate in the manufacture of selective weedkiller.

TABLE 18

| | Phenol | Ethanol |
|---|---|---|
| **Reaction with sodium** | a solution of phenol in dry ether liberates hydrogen on reaction with sodium: $2C_6H_5OH + 2Na$ $\rightarrow 2C_6H_5ONa + H_2$ | ethanol reacts with sodium liberating hydrogen: $2C_2H_5OH + 2Na$ $\rightarrow 2C_2H_5ONa + H_2$ |
| **Reaction with sodium hydroxide** | phenol is slightly acidic and dissolves to form sodium phenoxide $C_6H_5OH + NaOH$ $\rightarrow C_6H_5ONa + H_2O$ | no reaction |
| **Reaction with phosphorus pentachloride** | very slow reaction; products indicate presence of hydroxyl group: $C_6H_5OH + PCl_5$ $\rightarrow C_6H_5Cl + POCl_3 + HCl$ | vigorous reaction; products indicate presence of hydroxyl group: $C_2H_5OH + PCl_5$ $\rightarrow C_2H_5Cl + POCl_3 + HCl$ |
| **Ether formation** | sodium phenoxide reacts with halogeno-alkanes to produce phenyl ethers: $C_6H_5ONa + RBr$ $\rightarrow NaBr + C_6H_5—O—R$ | sodium ethoxide reacts with halogeno-alkanes to form ethers: $C_2H_5ONa + RBr$ $\rightarrow NaBr + C_2H_5—O—R$ |
| **Oxidation** | oxidation leads to polymerization | oxidized easily to form ethanal: $C_2H_5OH + (O)$ $\rightarrow H_2O + CH_3CHO$ |
| **Esterification** | reacts with acid chlorides to form esters: $C_6H_5OH + RCOCl$ $\rightarrow RCOOC_6H_5 + HCl$ | reacts with acid chlorides to form esters: $C_2H_5OH + RCOCl$ $\rightarrow RCOOC_2H_5 + HCl$ |

114

(e) The *Reimer–Tiemann reaction*: on refluxing a solution of phenol in sodium hydroxide with trichloromethane, salicylaldehyde is formed:

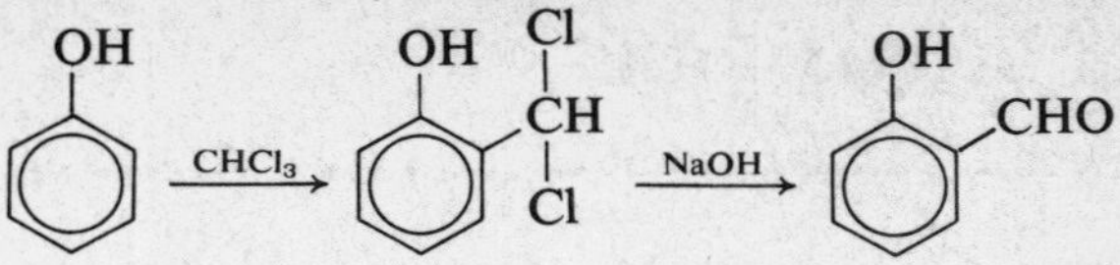

3 Reactions of the hydroxyl group: a comparison of the reactions given by phenol and an aliphatic alcohol are given in Table 18.

**Uses** Phenol is widely used as a disinfectant, although the use of phenol in this way is being supplanted by chlorinated phenols and similar derivatives. A large proportion of phenol is used in the plastics industry by way of the phenol-formaldehyde resins. It is also used in the production of dyes and explosives.

# 13 Nitro-Compounds and Phenylamine

## NITRO-COMPOUNDS

These compounds are produced when at least one hydrogen atom in the benzene nucleus has been replaced by the nitro ($-NO_2$) group. The simplest nitro-compound is nitrobenzene,

$$\text{C}_6\text{H}_5\text{NO}_2$$

The functional group is the nitro group, although reactions associated with the aromatic nucleus also take place.

### Preparation

Nitro-compounds are usually made by direct nitration. Benzene is nitrated using a mixture of concentrated nitric and sulphuric acids (nitrating acid) the main product being nitrobenzene if the temperature is kept below 60°C, but some m-dinitrobenzene is formed if the reaction is carried out above this temperature:

$$\text{C}_6\text{H}_6 + HNO_3 \xrightarrow{60°C} H_2O + \text{C}_6\text{H}_5NO_2 \xrightarrow[HNO_3]{100°C} \text{C}_6\text{H}_4(NO_2)_2$$

Fuming nitric acid produces m-dinitrobenzene. Methylbenzene nitrates under similar conditions to give a mixture of 2- and 4-nitro-derivatives.

### Properties and Reactions of Nitrobenzene

Nitrobenzene is a pale yellow liquid, with an almond-like odour. It is poisonous, immiscible with water but volatile in steam, while m-dinitrobenzene is a pale-yellow solid. Nitrobenzene is readily absorbed through the skin and the vapour is toxic.

The chief reactions of nitrobenzene are the reduction of the nitro group and the further nitration of the benzene nucleus.

1 Reduction of the nitro group. The main reaction is used in the preparation of phenylamine, which is produced in the laboratory by the reduction of nitrobenzene using tin and concentrated hydrochloric acid:

$$C_6H_5NO_2 + 6(H) \rightarrow 2H_2O + C_6H_5NH_2$$

By reducing nitrobenzene under different conditions (e.g. with zinc dust and ammonium chloride in neutral solution) different products are obtained, for example:

$$C_6H_5NO_2 + 4(H) \rightarrow C_6H_5NHOH + H_2O$$
phenyl hydroxylamine

2 Further nitration of nitrobenzene using nitrating acid or fuming nitric acid above 60°C produces m-dinitrobenzene, while prolonged nitration at temperatures in excess of 100°C yields 1,3,5-trinitrobenzene:

Nitrobenzene can be sulphonated and halogenated but it is unaffected by oxidizing and hydrolysing reagents.

## PHENYLAMINE (ANILINE)

Phenylamine is the simplest aromatic amine, derived by substituting one of the hydrogen atoms in ammonia by an aryl radical. Since phenylamine contains the —$NH_2$ group, it is a primary amine.

### Preparation

1 By the reduction of nitrobenzene. In the laboratory, the reduction is brought about using tin and concentrated hydrochloric acid:

$$2C_6H_5NO_2 + 3Sn + 12HCl$$
$$\rightarrow 2C_6H_5NH_2 + 4H_2O + 3SnCl_4$$

117

The phenylamine formed in this reaction forms a complex salt with the tin (IV) chloride:

$$SnCl_4 + 2HCl + 2C_6H_5NH_2 \rightarrow (C_6H_5NH_3^+)_2SnCl_6{}^{2-}$$

Phenylamine is liberated from this mixture by the addition of sodium hydroxide, and it is extracted by steam distillation. It is finally isolated from the distillate by ether extraction and purified by redistillation.

A similar process is used for manufacturing phenylamine. Nitrobenzene is reduced using iron turnings and dilute hydrochloric acid. After the reduction, calcium hydroxide is added, and the phenylamine is again recovered by steam distillation.

2 Phenylamine is also produced from benzamide on treatment with bromine water in alkaline solution:

$$C_6H_5CONH_2 + Br_2 + 4NaOH$$
$$\rightarrow C_6H_5NH_2 + 2NaBr + Na_2CO_3 + 2H_2O$$

## Properties and Reactions of Phenylamine

Phenylamine is a colourless liquid which turns brown on standing. It has a sickly, oily odour and it is poisonous.

Although a primary amine, phenylamine shows few resemblances to the aliphatic primary amines. It undergoes the usual reactions of the aromatic nucleus, and many special reactions associated with the —$NH_2$ group. In this way, phenylamine is an important intermediate in the preparation of other aromatic compounds.

1 Phenylamine is basic and, like the aliphatic amines, forms salts:

$$C_6H_5NH_2 + HCl \rightarrow C_6H_5NH_3{}^+Cl^-$$

phenylammonium<br>chloride

The free base is liberated from phenylammonium chloride (an off-white crystalline compound), by reaction with sodium hydroxide:

$$C_6H_5NH_3{}^+Cl^- + NaOH \rightarrow C_6H_5NH_2 + H_2O + Na^+Cl^-$$

2 In common with other primary amines, phenylamine produces an iso-nitrile on boiling with alkali and trichloromethane:

$$C_6H_5NH_2 + CHCl_3 + 3KOH \rightarrow 3H_2O + 3KCl + C_6H_5NC$$

The product, phenyl isonitrile is quickly recognized by its very objectionable odour. This test is characteristic of primary amines.

3 Phenylamine gives an intense purple colour with a suspension of bleaching powder in water. This colour reaction is often used as a test for phenylamine.

4 The reactions of the aromatic nucleus are as follows.

(a) When excess bromine water is added to a solution of phenylamine in hydrochloric acid, a precipitate of 2,4,6-tribromophenylamine is formed:

A similar reaction takes place with chlorine water.

(b) Sulphonation. On heating a solution of phenylamine in sulphuric acid, with fuming sulphuric acid, sulphanilic acid is produced:

(Sulphanilic acid contains both a basic and an acidic group in the same molecule.)

(c) The direct nitration of phenylamine is not possible since it is easily oxidized. Hence, nitro-derivatives of phenylamine are made by indirect methods.

5 The reactions of the —$NH_2$ group are as follows.

(a) Phenylamine reacts readily with ethanoyl chloride or ethanoic anhydride to form N-ethanoyl phenylamine:

N*-ethanoyl phenylamine may be hydrolysed by refluxing with acid, for example, 70% $H_2SO_4$:

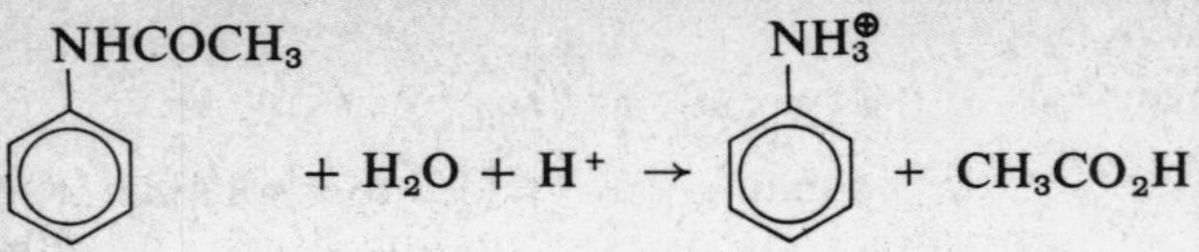

$$+ H_2O + H^+ \rightarrow \qquad + CH_3CO_2H$$

Phenylamine reacts with benzoyl chloride to form N*-benzoyl phenylamine

$$C_6H_5NH_2 + C_6H_5COCl \rightarrow HCl + C_6H_5CONHC_6H_5$$

(b) Phenylamine reacts with halogeno-alkanes by heating under pressure. Secondary or tertiary amines are produced according to the proportion of halogeno-alkane used.

(c) A special reaction, known as *diazotization*, takes place when phenylamine (or any primary aromatic amine) is treated with ice-cold nitrous acid (formed by mixing sodium nitrite and ice-cold hydrochloric acid). The product is benzenediazonium chloride, and the molecule contains two adjacent nitrogen atoms linked by a triple bond:

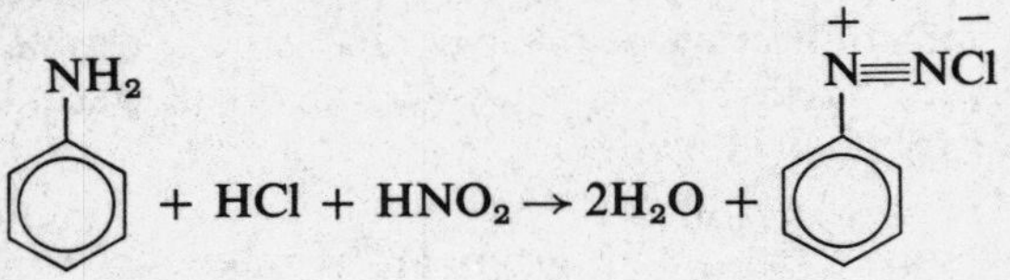

Diazonium compounds are stable only below 5°C, and they are highly reactive. Consequently, they are important in the production of other aromatic compounds, and the reactions of benzenediazonium chloride are outlined below.

* The symbol N indicates that the ethanoyl or benzoyl group is linked to a nitrogen atom and not to a carbon atom of the parent compound.

## Reactions of Benzenediazonium Chloride

1 It is reduced to phenylhydrazine using tin and hydrochloric acid:

$$C_6H_5N_2Cl + 4(H) \rightarrow HCl + C_6H_5NH.NH_2$$

2 With primary amines, for example phenylamine, diazoamino-benzene (or a derivative) is formed:

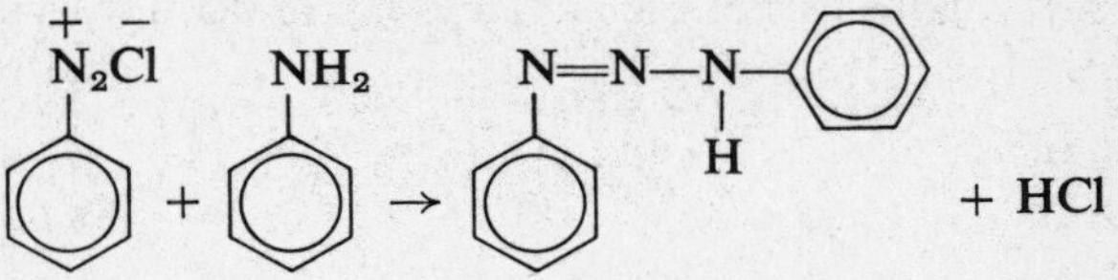

3 A solution of benzene diazonium chloride reacts with an alkaline solution of phenol in a manner known as *coupling*:

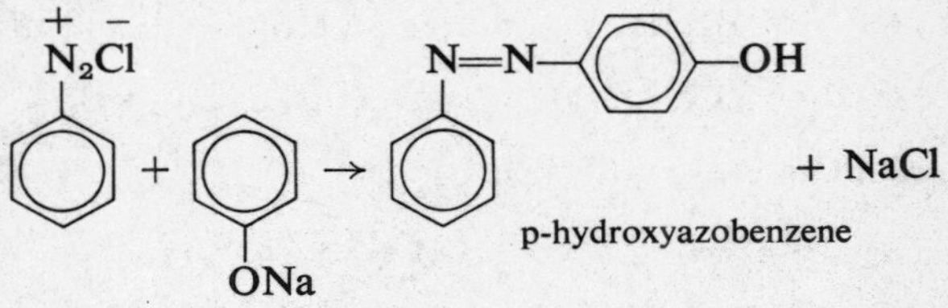

p-hydroxyazobenzene

Many of the compounds formed by coupling reactions are highly coloured.

4 On heating with ethanol, benzene is produced:

$$C_6H_5N_2Cl + CH_3CH_2OH \rightarrow C_6H_6 + CH_3CHO + N_2 + HCl$$

5 On boiling with a little dilute acid, phenol is produced from benzene diazonium chloride:

$$C_6H_5N_2Cl + H_2O \rightarrow C_6H_5OH + HCl + N_2$$

6 *Sandmeyers reaction.* In these reactions, which are usually catalysed by copper (I) salts, the diazo-group is replaced by a halogen or a nitrile group. For example:

(a) $C_6H_5N_2Cl \xrightarrow[\text{in conc. HCl}]{\text{CuCl}} C_6H_5Cl + N_2$

$\xrightarrow[\text{HCl}]{\text{CuBr in}}$

$C_6H_5Br + N_2$

(b) $C_6H_5N_2Cl + KI \xrightarrow[\text{aqueous solution}]{\text{heat in}} KCl + N_2 + C_6H_5I$

(c) $C_6H_5N_2Cl + CuCN \xrightarrow{KCN} N_2 + CuCl + C_6H_5CN$

*Note:* that in reactions 4, 5 and 6, the diazo-group is replaced and nitrogen is evolved.

## Phenylmethylamine

This aromatic amine carries the amino group in the side chain.

$$CH_2.NH_2$$

It is prepared:

1 By the action of alcoholic ammonia on (chloromethyl) benzene:

$$C_6H_5CH_2Cl + NH_3 \rightarrow HCl + C_6H_5CH_2NH_2$$

2 By the reduction of benzonitrile:

$$C_6H_5CN + 4(H) \rightarrow C_6H_5CH_2NH_2$$

It would be expected that phenylmethylamine should show a close resemblance to the aliphatic amines in those reactions involving the —$NH_2$ group. This is found to be true, and the chief properties of ethylamine, phenylmethylamine and phenylamine are compared in Table 19.

## The Relative Basic Strength of Amines

The strength of a base may be expressed in terms of the dissociation constant $K_b$ (see Part 3), or as a $pK_b$ value (where $pK_b = -\log K_b$).

$pK_b$ values for ammonia, methylamine, ethylamine and phenylamine are 4·76, 3·36, 3·33 and 9·38 respectively. We would expect methylamine and ethylamine to be stronger bases than ammonia, since the $+I$ (electron repelling) inductive effect of the alkyl group allows the lone pair on the nitrogen atom to accept a proton more readily.

$$CH_3 \rightarrow \overset{\displaystyle H}{\underset{\displaystyle H}{N:}}$$

Phenylamine is a weaker base than ammonia since stability is gained by the lone pair of electrons becoming part of the delocalized system, and thus less available for donation to a proton.

TABLE 19

| | Ethylamine | Phenylmethylamine | Phenylamine |
|---|---|---|---|
| Formula | $CH_3CH_2NH_2$ | $C_6H_5CH_2NH_2$ | $C_6H_5NH_2$ |
| Physical properties | liquid with a fishy-ammoniacal odour; soluble in water | liquid with ammoniacal odour; sparingly soluble in water | liquid with a sickly, oily smell; sparingly soluble in water |
| Basic properties | strongly basic; solution in water is alkaline, turns litmus blue, forms salts and precipitates metal hydroxides | basic; forms salts with acids | weak base; forms salts with acids |
| Sulphonation | forms a salt; does not sulphonate | aromatic nucleus undergoes sulphonation | aromatic nucleus undergoes sulphonation |
| Reaction with ice-cold nitrous acid | liberates nitrogen and forms $CH_3CH_2OH$ (ethanol) | liberates nitrogen and forms $C_6H_5CH_2OH$ (phenylmethanol) | does not liberate nitrogen, but diazotizes to form $C_6H_5N_2Cl$ (benzene diazonium chloride) |

# 14   Aromatic Aldehydes, Ketones and Acids

## BENZALDEHYDE

In benzaldehyde (phenylmethanal) the aldehyde group is directly attached to the aromatic nucleus.

### Preparation

1 By oxidation of phenylmethanol (benzyl alcohol):

$$C_6H_5CH_2OH + (O) \rightarrow C_6H_5CHO + H_2O$$

2 By hydrolysis of benzal chloride:

$$C_6H_5CHCl_2 + 2NaOH \rightarrow C_6H_5CHO + 2NaCl + H_2O$$

3 By reduction of benzoyl chloride using hydrogen over a partially poisoned palladium catalyst:

$$C_6H_5COCl + H_2 \rightarrow HCl + C_6H_5CHO$$

Benzaldehyde is used in perfumes, and in the manufacture of flavourings and dyes. It is produced industrially by the hydrolysis of benzal chloride or by the vapour-phase oxidation of methylbenzene.

### Properties and Reactions of Benzaldehyde

Benzaldehyde is a colourless liquid having a strong almond-like odour. It is sparingly soluble in water.

Benzaldehyde shows a considerable resemblance to ethanal although the presence of the aromatic nucleus gives rise to a number of differences.

1 Benzaldehyde undergoes the usual substitution reactions asso-

124

ciated with aromatic compounds, although it reacts less readily than benzene:

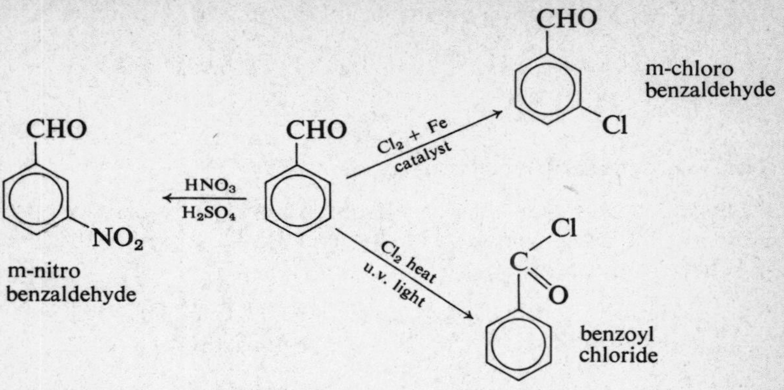

2 Oxidation of benzaldehyde is effected by mild oxidizing agents, for example dilute nitric acid. (The concentrated acid appears to nitrate the nucleus rather than oxidize the side chain.)

$$C_6H_5CHO + (O) \rightarrow C_6H_5CO_2H$$

Atmospheric oxygen will oxidize benzaldehyde, and the white crystals deposited round the stopper of a partly used bottle of benzaldehyde are benzoic acid formed by atmospheric oxidation. However, benzaldehyde does not give any significant reduction of Fehling's solution or form a silver mirror.

3 The reactions of the carbonyl group are similar to those given by the aliphatic aldehydes.

(a) Reduction by sodium amalgam produces phenylmethanol:

$$C_6H_5CHO + H_2 \rightarrow C_6H_5CH_2OH$$

(b) Benzaldehyde undergoes addition reactions with sodium hydrogen sulphite and hydrogen cyanide:

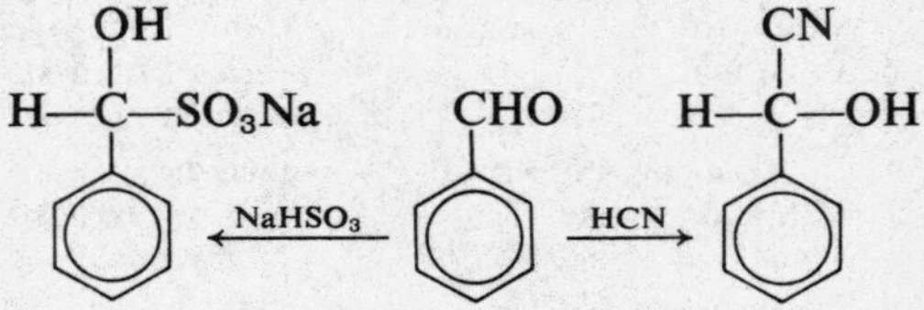

With ammonia, a complex substance is formed:

$$3C_6H_5CHO + 2NH_3 \rightarrow {}_3H_2O + (C_6H_5CH:N)_2CHC_6H_5$$

(c) Benzaldehyde does not undergo polymerization.

(d) Benzaldehyde takes part in the same type of condensation reactions shown by the aliphatic aldehydes. For example, the reaction between benzaldehyde and hydroxylamine is

$$C_6H_5CHO + H_2NOH \rightarrow H_2O + C_6H_5CH{=}NOH$$
benzaldoxime

## Special Reactions of Benzaldehyde

1 Benzaldehyde reacts with an alkaline solution of ethanal to produce cinnamaldehyde (3-phenylpropanal). This is known as the *Claisen* reaction.

$$C_6H_5CHO + CH_3CHO \rightarrow H_2O + C_6H_5CH{=}CHCHO$$
cinnamaldehyde

TABLE 20

|  | *Ethanal* | *Benzaldehyde* |
|---|---|---|
| *Formula* | $CH_3CHO$ | $C_6H_5CHO$ |
| *Physical properties* | colourless, volatile liquid, soluble in water | colourless, oily liquid, immiscible with water |
| *Reducing properties* | readily reduces Fehling's solution and forms a silver mirror | does not reduce Fehling's solution; forms a slight silver mirror |
| *Further oxidation* | easily oxidized to ethanoic acid | easily oxidized to benzoic acid |
| *Haloform reaction using iodine* | forms tri-iodomethane (iodoform) | no reaction |
| *Substitution* | chlorine substitutes alkyl group hydrogens to give trichlorethanal | gives the usual substitution products associated with aromatic compounds |
| *Addition reactions* | adds on hydrogen cyanide, sodium hydrogen sulphite and ammonia | adds on hydrogen cyanide, sodium hydrogen sulphite; forms a complex compound with ammonia |
| *Condensation reactions* | gives condensation products with hydroxylamine, phenylhydrazine, etc. | gives condensation products with hydroxylamine, phenylhydrazine, etc. |
| *Schiff's reagent* | colour produced | colour produced |
| *Action of alkali* | resinifies on heating | gives the Cannizzaro reaction |
| *Polymerization* | forms many polymers | does not polymerize |

2 Benzaldehyde undergoes the Cannizzaro reaction on shaking with a solution of potassium or sodium hydroxide at room temperature:

$$2C_6H_5CHO + NaOH \rightarrow C_6H_5CO_2Na + C_6H_5CH_2OH$$

A comparison of the properties of ethanal and benzaldehyde is given in Table 20.

## ACETOPHENONE AND BENZOPHENONE

These ketones have the structural formulae:

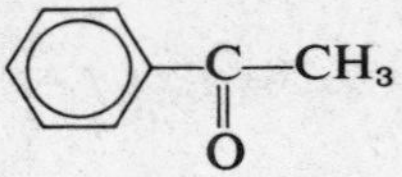

acetophenone
(phenylethanone)

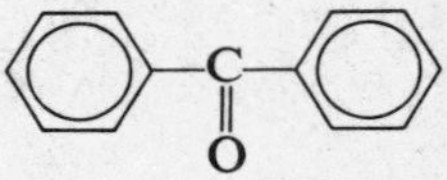

benzophenone

Both of these ketones are produced by the Friedel–Crafts reaction:

$$C_6H_6 + CH_3COCl \xrightarrow{\text{AlCl}_3} C_6H_5CO.CH_3 + HCl$$

$$C_6H_6 + C_6H_5COCl \xrightarrow{\text{AlCl}_3} HCl + C_6H_5CO.C_6H_5$$

TABLE 21

|  | Propanone | Acetophenone | Benzophenone |
|---|---|---|---|
| Formula | $(CH_3)_2CO$ | $C_6H_5COCH_3$ | $(C_6H_5)_2CO$ |
| Physical properties | liquid with a strong, pleasant odour | colourless solid (m.p. 20°C) with an almond-like odour | colourless solid with pleasant aromatic odour |
| Reducing properties | none | none | none |
| Haloform reaction using iodine | forms tri-iodomethane (iodoform) | forms tri-iodomethane (iodoform) | no reaction |
| Substitution reactions | alkyl hydrogen atoms substituted by chlorine | substitution (e.g. nitration) takes place in aromatic nucleus | substitution takes place in the aromatic nucleus |
| Addition reactions | adds on hydrogen cyanide and sodium hydrogen sulphite | addition products not formed | addition products not formed |

Both compounds are solids, and their reactions resemble those of aliphatic ketones. The chief properties of propanone, acetophenone and benzophenone are compared in Table 21.

## BENZOIC ACID (BENZENECARBOXYLIC ACID)

In aromatic carboxylic acids, the carboxyl group is attached directly to the aromatic nucleus. Benzoic acid is the simplest aromatic monocarboxylic acid; it has the following structural formula:

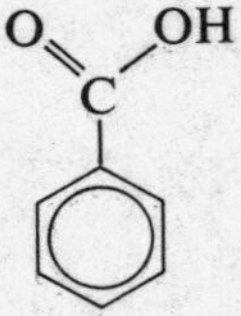

### Preparation

1 From benzaldehyde by oxidation:

$$C_6H_5CHO + (O) \rightarrow C_6H_5CO_2H$$

2 By the hydrolysis of benzoates:

$$C_6H_5COOR + NaOH \rightarrow C_6H_5CO_2Na + ROH$$

Benzoic acid is liberated on acidification of the reaction mixture.

3 By hydrolysing benzonitrile in the presence of acids or bases:

$$C_6H_5CN + 2H_2O + H^+ \rightarrow C_6H_5CO_2H + NH_4^+$$

4 By the oxidation of methylbenzene using potassium permanganate:

$$C_6H_5CH_3 + 3(O) \rightarrow C_6H_5CO_2H + H_2O$$

### Properties and Reactions of Benzoic Acid

Benzoic acid is a white, crystalline solid with a faint odour. It sublimes on heating, and is sparingly soluble in cold water, but more soluble in hot water.

The reactions of the carboxyl group are similar to those of the aliphatic acids. In addition, benzoic acid shows the usual properties associated with aromatic compounds. The reactions may be summarized as follows:

128

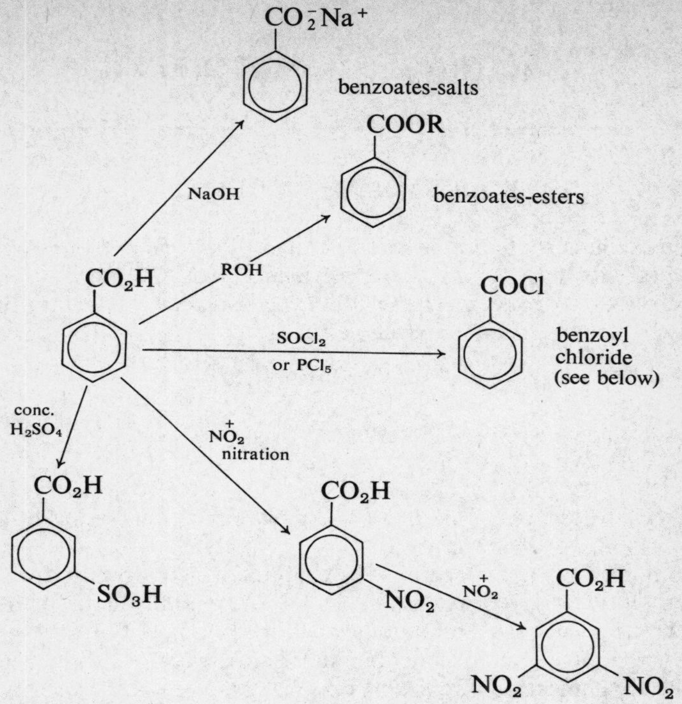

CO₂⁻Na⁺ → benzoates-salts

COOR → benzoates-esters

COCl → benzoyl chloride (see below)

CO₂H

NaOH

ROH

SOCl₂ or PCl₅

conc. H₂SO₄

NO₂⁺ nitration

CO₂H / SO₃H

CO₂H / NO₂

NO₂⁺

CO₂H / NO₂ NO₂

The reactions of benzoyl chloride are as follows:

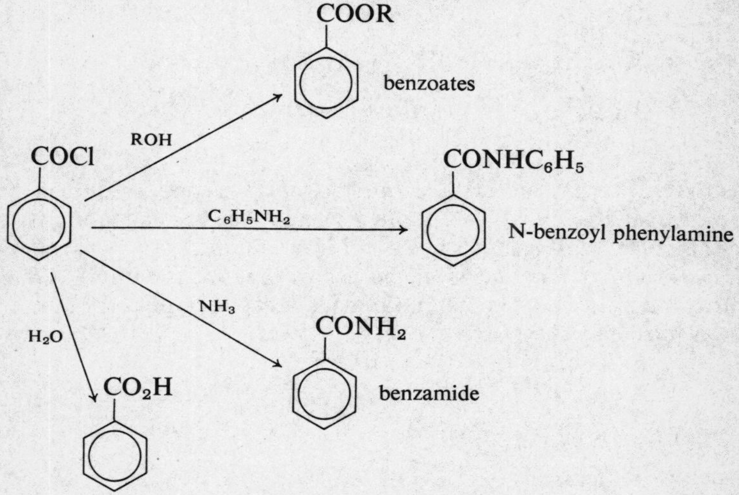

COCl

ROH → COOR benzoates

C₆H₅NH₂ → CONHC₆H₅ N-benzoyl phenylamine

NH₃ → CONH₂ benzamide

H₂O → CO₂H

# 15 Bifunctional Compounds

## POLYHYDRIC ALCOHOLS

These alcohols contain at least two hydroxyl groups in the molecule. *Dihydric* alcohols, for example ethylene glycol, $CH_2OH.CH_2OH$, contain two hydroxyl groups, while *trihydric* alcohols contain three hydroxyl groups in the molecule, as in glycerol:

$$CH_2OH$$
$$|$$
$$CHOH$$
$$|$$
$$CH_2OH$$

The IUPAC names for these alcohols are ethane-1,2-diol and propane-1,2,3-triol respectively.

Both alcohols are colourless, oily liquids which are miscible with water. Ethylene glycol is used as an antifreeze, and the preparation and reactions of this compound are outlined on p. 161.

Glycerol, which is hygroscopic and tastes sweet, is manufactured by the saponification of oils and fats which are esters of glycerol and long-chain carboxylic acids:

$$\begin{array}{lll} R.CO_2CH_2 & CH_2OH & R.CO_2Na \\ | & | & \\ R'.CO_2CH + 3NaOH \rightarrow CHOH + & R'.CO_2Na \\ | & | & \\ R''.CO_2CH_2 & CH_2OH & R''.CO_2Na \end{array} \Bigg\} \text{ soap}$$

oil or fat

Glycerol shows the usual properties of an alcohol, although the presence of three hydroxyl groups means that it is capable of reacting in three stages. It reacts with sodium, phosphorus pentachloride and organic acids in the usual way. A mixture of fuming nitric and concentrated sulphuric acids react with glycerol to form the trinitrate at 10°C:

$$\begin{array}{lll} CH_2OH & & CH_2ONO_2 \\ | & & | \\ CHOH + 3HNO_3 \rightarrow 3H_2O + & CHONO_2 \\ | & & | \\ CH_2OH & & CH_2ONO_2 \end{array}$$

Glyceryl trinitrate (nitroglycerin) is an oily liquid which explodes when heated quickly, or when struck. It forms the explosive constituent of dynamite, gelignite, etc.

**Reactions of Ethane-1,2-diol**

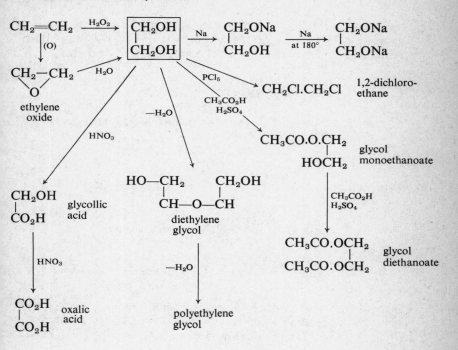

## VINYL COMPOUNDS

Addition of a hydrogen halide to ethyne produces a vinyl halide:

$$CH{\equiv}CH + HX \rightarrow CH_2{=}CHX$$

Unexpectedly, the halogen atom in vinyl halides is less reactive than in halogeno-alkanes—for example, they do not form Grignard reagents.

Vinyl ethanoate is made by passing ethanoic acid vapour and ethyne over a zinc ethanoate catalyst:

$$CH{\equiv}CH + CH_3CO_2H \rightarrow CH_3CO_2CH{=}CH_2$$

131

Cyanoethene (acrylonitrile) is made by the action of hydrogen cyanide on ethyne:

$$CH{\equiv}CH + HCN \rightarrow CH_2{=}CHCN$$

The importance of all these vinyl compounds is in the production of polymeric vinyl resins; for example, vinyl chloride polymerizes to PVC, while polyvinyl acetate is used widely in surface coatings, and in emulsion paints. The polymerization of a mixture of vinyl compounds results in the formation of a co-polymer resin which again is used in the manufacture of paints and surface coatings.

## DICARBOXYLIC ACIDS

The simplest dicarboxylic acid is ethanedioic acid (oxalic acid),

$$\begin{array}{c} CO_2H \\ | \\ CO_2H \end{array}$$

It is prepared by the hydrolysis of cyanogen or by heating sodium methanoate:

(a)  $\begin{array}{c} CN \\ | \\ CN \end{array} + 4H_2O \rightarrow \begin{array}{c} CO_2NH_4 \\ | \\ CO_2NH_4 \end{array}$  ammonium oxalate

(b)  $2HCO_2Na \xrightarrow{\text{heat}} H_2 + (CO_2Na)_2 \xrightarrow{Ca^{2+}} (CO_2)_2Ca$
(precipitated)

In both cases, the acid is obtained by acidification of the reaction mixture. Ethanedioic acid is a white, crystalline solid which is soluble in water and produces an acid solution. It is poisonous.

### Reactions of Ethanedioic Acid

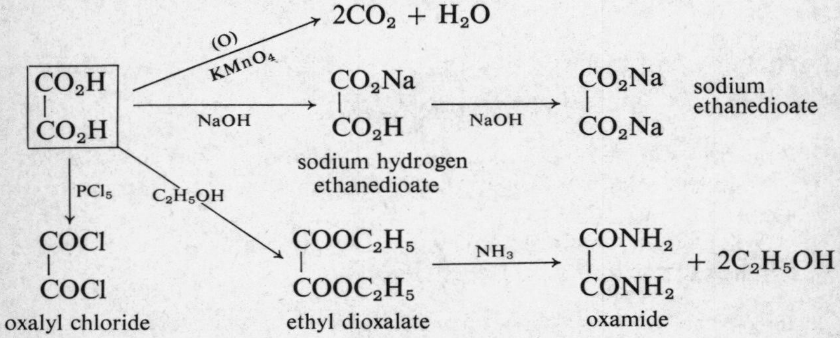

$$\begin{array}{c} CO_2H \\ | \\ CO_2H \end{array} \xrightarrow[\text{KMnO}_4]{(O)} 2CO_2 + H_2O$$

$$\begin{array}{c} CO_2H \\ | \\ CO_2H \end{array} \xrightarrow{\text{NaOH}} \begin{array}{c} CO_2Na \\ | \\ CO_2H \end{array} \xrightarrow{\text{NaOH}} \begin{array}{c} CO_2Na \\ | \\ CO_2Na \end{array} \text{ sodium ethanedioate}$$

sodium hydrogen ethanedioate

$$\begin{array}{c} \downarrow PCl_5 \qquad C_2H_5OH \end{array}$$

$$\begin{array}{c} COCl \\ | \\ COCl \end{array} \qquad \begin{array}{c} COOC_2H_5 \\ | \\ COOC_2H_5 \end{array} \xrightarrow{NH_3} \begin{array}{c} CONH_2 \\ | \\ CONH_2 \end{array} + 2C_2H_5OH$$

oxalyl chloride        ethyl dioxalate        oxamide

Other dicarboxylic acids include:

$$\underset{\substack{\\ \text{malonic or} \\ \text{propane-1,3-dioic acid}}}{\overset{\displaystyle \text{CO}_2\text{H}}{\underset{\displaystyle \text{CO}_2\text{H}}{\text{CH}_2}}}$$

$$\underset{\substack{\\ \text{succinic or} \\ \text{butane-1,4-dioic acid.}}}{\overset{\displaystyle \text{CH}_2\text{CO}_2\text{H}}{\displaystyle \text{CH}_2\text{CO}_2\text{H}}}$$

## AMINOETHANOIC ACID (GLYCINE)

Aminoethanoic acid, $CH_2NH_2CO_2H$, is the simplest amino acid. It is prepared by reacting monochloroethanoic acid with concentrated ammonia:

$$CH_2ClCO_2H + NH_3 \rightarrow CH_2(NH_2)CO_2H + HCl$$

Aminoethanoic acid contains both an acidic and a basic group in the same molecule; hence it reacts with

(a) alkalies to form salts:

$$CH_2NH_2CO_2H + NaOH \rightarrow H_2O + CH_2NH_2CO_2{}^-Na^+$$

(b) acids, also forming salts:

$$CH_2NH_2CO_2H + HCl \rightarrow \underset{\displaystyle CH_2CO_2H}{NH_3^+Cl^-}$$

A proton from its own carboxyl group can migrate to the amino group, forming an ion which carries both a positive and a negative charge. This is known as a *zwitterion*.

$$\underset{\displaystyle NH_2}{CH_2CO_2H} \rightleftharpoons \underset{\displaystyle NH_3^+}{CH_2CO_2^-}$$

## Reactions of Glycine

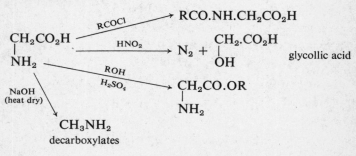

$$\underset{\displaystyle NH_2}{CH_2CO_2H}$$

RCOCl → $RCO.NH.CH_2CO_2H$

HNO$_2$ → $N_2 + \underset{\displaystyle OH}{CH_2.CO_2H}$ glycollic acid

ROH / H$_2$SO$_4$ → $\underset{\displaystyle NH_2}{CH_2CO.OR}$

NaOH (heat dry) → $CH_3NH_2$ decarboxylates

133

# UREA

Urea is a colourless solid, which is soluble in water. It has the structural formula

$$\begin{array}{c} NH_2 \\ | \\ C=O \\ | \\ NH_2 \end{array}$$

and its reactions are essentially those of the amino group rather than the carbonyl group. Urea is used as a nitrogenous fertilizer and in the manufacture of plastics.

## Preparation

1 By reacting carbonyl chloride with ammonia:

$$COCl_2 + 2NH_3 \rightarrow 2HCl + CO(NH_2)_2$$

2 It is manufactured by heating carbon dioxide with excess ammonia to 200°C and at 200 atmospheres pressure. Ammonium carbamate is formed as an intermediate.

$$CO_2 + 2NH_3 \rightleftharpoons \begin{array}{c} O^-(NH_4^+) \\ | \\ C=O \\ | \\ NH_2 \end{array} \rightleftharpoons CO(NH_2)_2 + H_2O$$

## Reactions of Urea

1 On gentle heating, biuret is formed:

$$2CO(NH_2)_2 \rightarrow NH_2CONH.CO.NH_2 + NH_3$$

2 Urea is basic, and forms salts with acids:

$$CO(NH_2)_2 + HNO_3 \rightarrow CO(NH_2)_2H^+NO_3^-$$

Urea functions as a monoacidic base only.

3 It is hydrolysed by acids or alkalies:

$$CO(NH_2)_2 + 2NaOH \rightarrow 2NH_3 + Na_2CO_3$$

$$CO(NH_2)_2 + 2HCl + H_2O \rightarrow CO_2 + 2NH_4Cl$$

4 With nitrous acid, nitrogen is liberated and carbon dioxide is formed:

$$CO(NH_2)_2 + 2HNO_2 \rightarrow CO_2 + 2N_2 + 2H_2O$$

# RING CLOSURE

When the two groups in a bifunctional compound react with each other, two possibilities arise.

1 Reaction may take place between two molecules—this is an *intermolecular* reaction and it usually leads to the formation of long-chain polymers (see Chapters 17 and 18).
2 Reaction may take place between the two groups in the same molecule—this is an *intramolecular* reaction and leads to the formation of a cyclic compound. Some examples of ring closure are given below.

$$\begin{array}{l} CH_2CH_2CO_2H \\ | \\ CH_2CH_2CO_2H \end{array} \xrightarrow[\text{anhydride}]{\text{distil with ethanoic}} \begin{array}{l} CH_2\!-\!CH_2 \\ | \qquad\qquad CO \\ CH_2\!-\!CH_2 \end{array} + CO_2 + H_2O$$

hexane-1,6-dioic acid          cyclopentanone

benzene-1,
2-dicarboxylic acid
(phthalic acid)

benzene-1,2-dicarboxylic
anhydride (phthalic anhydride)

4-hydroxy-butanoic
acid

lactone

The last example is an instance of intramolecular esterification. Ring closure is governed by:

(a) the number of atoms present in the ring (strain factor). When fewer than five atoms are present in the cyclic structure, the bonds between neighbouring atoms are pulled away from their normal angles. The resulting strain reduces the chance of ring closure. At the other end of the scale, very long-chain bifunctional compounds do not form rings easily, as the two reactive ends of the molecule are unlikely to come close enough for reaction to take place.
(b) size of the atoms and groups. Bulky groups present in the molecule can obviously act as a barrier to ring closure.

# 16 Isomerism

Isomerism is the term denoting the fact that a certain molecular formula may correspond to more than one compound. This phenomenon arises because it is possible for the atoms in an organic molecule to be joined together in different ways; that is, the molecules are structurally different. Isomerism may be divided into two broad classes, structural isomerism and stereoisomerism.

**Structural Isomerism**

The same atoms are joined in a different sequence. Table 22 shows some further subdivision of this type of isomerism.

TABLE 22 *Classes of structural isomerism*

| Class | Characteristics | Examples |
|---|---|---|
| (a) *Positional isomerism* | The groups are attached at different positions in a carbon chain or aromatic nucleus. Isomers have same functional group and show similar chemical reactions | (i) $CH_3.CH.CH_3$ with $OH$ and $CH_3CH_2CH_2OH$ (ii) |
| (b) *Metamerism* | The two groups attached to a central functional group differ. Isomers have same functional group, and show very similar chemical behaviour. | $C_2H_5$ $>NH$ $C_2H_5$ and $CH_3$ $>NH$ $C_3H_7$ |
| (c) *Functional group isomerism* | Functional group differs between two isomers. Hence chemical properties quite different | $CH_3CH_2OH$ and $CH_3—O—CH_3$ |

## Stereoisomerism

The same atoms are joined in the same sequence by bonds which point in different directions in space.

The four bonding pairs of electrons surrounding a four-covalent carbon atom would be expected to repel each other to positions as far apart as possible. This leads to a tetrahedral arrangement of the bonds round a carbon atom, and this is in agreement with the fact that no isomers of the type

$$X-\underset{\underset{Y}{|}}{\overset{\overset{H}{|}}{C}}-H \qquad \left( X-\underset{\underset{H}{|}}{\overset{\overset{H}{|}}{C}}-Y \right)$$

have been discovered.

Three types of stereoisomerism have been recognized: optical, geometrical and conformational.

## Optical Isomerism

The existence of optical activity can be shown by the following experiments using polarized light.

A beam of monochromatic light (e.g. from a sodium lamp) which passes through a polarizing crystal (made from pure calcium carbonate or fluoride) is visible in the eyepiece of a polarimeter after

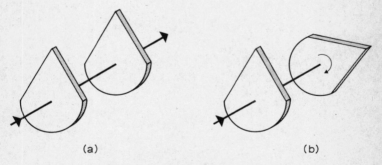

(a)                                    (b)

Fig. 13 Crystal alignment in a polarimeter

passing through a second crystal *only when the two crystals are identically aligned*. Should the analysing crystal (the second crystal) be rotated out of alignment, the intensity of the light seen in the eyepiece quickly diminishes.

Certain substances (usually present in solution) when interposed between the polarizing and analysing crystal, require the rotation of the analysing crystal either to the right or left (clockwise or anti-clockwise) in order to obtain maximum illumination in the eyepiece. Such substances are said to be *optically active* as they bring about a

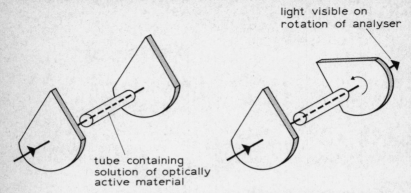

light visible on
rotation of analyser

tube containing
solution of optically
active material

Fig. 14 Rotation of the plane of polarized light by an optically
active substance

rotation of the plane of polarized light passing along the polarimeter tube.

Optically active substances are classified as *dextro-rotatory* when the analyser has to be turned clockwise, and *laevo-rotatory* when the

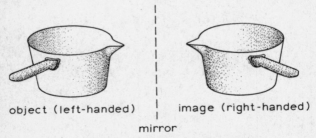

object (left-handed)     image (right-handed)

mirror

Fig. 15 Right- and left-handed structures

analyser requires turning anti-clockwise in order to obtain maximum illumination in the eyepiece.

Substances which exhibit optical activity are those for which it is possible to construct a right-handed and a left-handed structure which cannot be superimposed. Such structures are termed *asymmetrical*. (A simple example of this is a saucepan, which has a

138

pouring lip, thus making it right- or left-handed, as shown in Figure 15. These two structures are related to each other as object and mirror image.) Thus, a molecule having two non-superimposable structures which are related to each other as object and mirror image

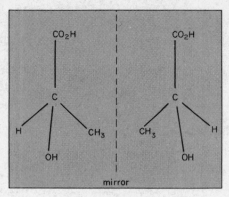

Fig. 16 Optical isomers of lactic acid

shows optical activity, one form being dextro- and the other laevo-rotatory. Lactic acid (2-hydroxy-propanoic acid) is an example of this type of compound, and the structures of the two optical isomers are shown in Figure 16.

Lactic acid is made from ethanal by way of hydrolysing the cyanhydrin:

$$CH_3CHO \xrightarrow{HCN} CH_3\!-\!\underset{\underset{H}{|}}{\overset{\overset{OH}{|}}{C}}\!-\!CN \xrightarrow{hydrolyse} CH_3\!-\!\underset{\underset{H}{|}}{\overset{\overset{OH}{|}}{C^*}}\!-\!COOH$$

The central carbon atom in the lactic acid molecule carries four different groups (or atoms) and it is said to be asymmetric. To indicate this, an asterisk is placed by this carbon atom when the formula for the molecule is written out.

The two forms of lactic acid, which differ only in the direction in which they rotate the plane of polarized light, are known as *enantiomorphs*. Normally, the two forms would be produced in equal quantities by the above method, giving a mixture in which no optical activity is apparent, since the rotation due to one form is exactly cancelled by that due to the other. Such a mixture is termed a *racemic mixture*. The separation of a racemic mixture into the two active components is called *resolution*.

139

A compound with two different asymmetric carbon atoms, that is two centres of activity, would be expected to have four optical isomers. An example of this is 1,2-dichloro-1-phenyl-propane:

$$C_6H_5.C^*H—C^*H.CH_3$$
$$\quad\quad\ \ \ |\quad\quad\ |$$
$$\quad\quad\ \ \ Cl\quad\quad Cl$$

Each centre of activity is capable of contributing to a different extent to the total rotatory power of the molecule. (For example, if the rotatory power of the first centre is 30° and the second 20°, then the rotatory powers of the four isomers will be $+50°$, $+10°$, $-10°$ and $-50°$.)

## Resolution of Racemic Mixtures

When an optically active substance is produced by a chemical reaction, it is usual to find that the d and l forms are produced in equal quantities. (In biological systems, this is very seldom the case, see Chapter 17.) A method of separation (resolution) of the two optical isomers of an acid present in a racemic mixture is to form a crystalline salt using either the pure d (or pure l) form of an optically active base.

$$(\text{Enantiomorphs})\begin{cases}\text{d-acid}\\\text{l-acid}\end{cases} + \text{d'-base} \rightarrow \begin{rcases}\text{dd'-salt}\\\text{ld'-salt}\end{rcases}(\text{diastereoisomers})$$

The two salts (called diastereoisomers) have different rotatory powers, different solubilities, melting points, etc. Therefore, it is possible to separate the two salts by fractional crystallization or large-scale chromatography. Once the salts have been obtained pure, treatment with dilute sodium hydroxide liberates the organic base which is removed, and subsequent acidification liberates the free acid. (The converse of this process can be applied to the resolution of the racemic mixture of an optically active base.)

## Geometrical Isomerism

When adjacent carbon atoms are joined by a double bond, rotation of one carbon atom relative to the other is impossible. This fixes the direction in space in which the two other bonds attached to each carbon atom can take, and a type of isomerism known as geometrical isomerism results.

For example, 1,2-dichloroethene, $ClCH\!=\!CHCl$, shows geometrical isomers in which

140

(a) both chlorine atoms are on the same side of the molecule—this is called the *cis* form.

(b) the chlorine atoms are on opposite sides of the molecule; this is the *trans* form. Although chemically similar, they show differences in physical properties: the *cis* isomer boils at 60°C, while the *trans* isomer has a boiling point of 49°C. These isomers are not optically active, but they may be readily distinguished since only the *cis* form has a dipole moment.

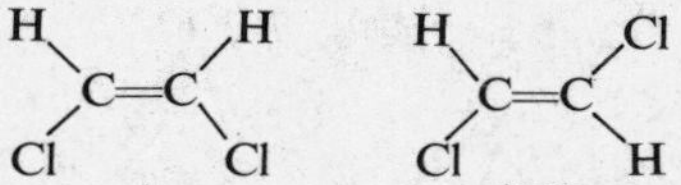

Fig. 17 Geometrical isomers of 1,2-dichloroethene

Maleic and fumaric acids form another example of geometrical isomerism:

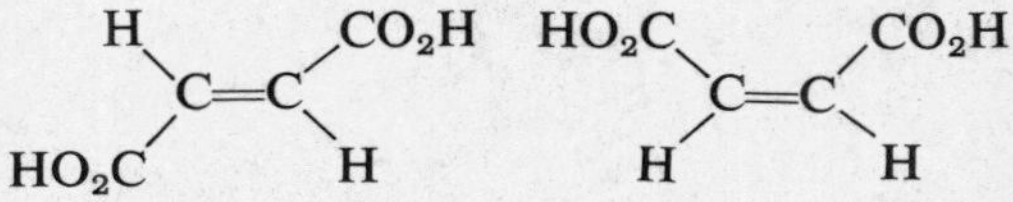

fumaric acid, m.p. 288°C          maleic acid, m.p. 130°C
    (*trans*-isomer)                (*cis*-isomer)

Maleic acid is converted into maleic anhydride on heating to 150°C whereas fumaric acid forms maleic anhydride on prolonged heating at 250°C.

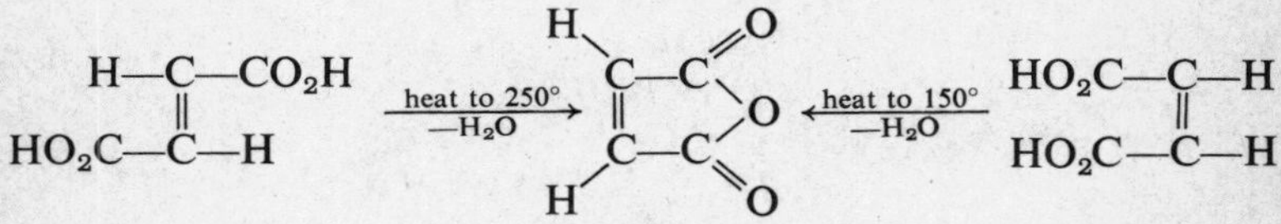

maleic anhydride

**Conformational Isomerism**

It is assumed that groups joined by single bonds are free to rotate; for example, in the ethane molecule the hydrogen atoms may take up any relative positions when the molecule is viewed from one end, along the carbon–carbon axis, as shown in Figure 18.

However, when bulky atoms or groups replace the hydrogen atoms in ethane, it is unlikely that the molecule would take the shape shown in Figure 18(a), rather the shape would be more like Figure 18(b). In

141

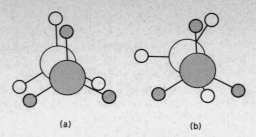

Fig. 18  Free rotation in the ethane molecule

a molecule of this shape, there is little strain in accommodating the bulky groups in the structure. Free rotation is still possible, but on balance, more molecules will have a shape similar to (b) than (a). Conformation relates to the preferred shapes adopted by a molecule and further discussion of this topic lies outside the scope of this book.

# 17   Biologically Important Compounds

The designation *organic* automatically links this aspect of chemistry with living organisms—or more generally, biological processes. Most of the chemicals encountered in a study of biological systems are very complicated molecules, but in this chapter, some of the properties of the simpler molecules which relate to living systems are described.

## Carbohydrates

This class of compounds are among the most common constituents of plants and animals. In animals they provide a source of energy, while in plants they form the greater part of the cell tissue on which the plant relies for its support. Vitamin C (ascorbic acid) is a carbohydrate and other carbohydrates have important effects on biological activity. The name originates from the fact that many sugars (which form part of this class of compounds) have the general formula $C_x(H_2O)_y$, that is carbo(n)–hydrate. Carbohydrates may be grouped as follows:

CARBOHYDRATES

    SUGARS (sweet, soluble, crystalline)

        MONOSACCHARIDES—pentoses and hexoses, e.g. ribose and glucose

        DISACCHARIDES e.g. sucrose (cane-sugar) and maltose

    POLYSACCHARIDES (tasteless, insoluble, amorphous) — starches, celluloses, etc.

**Monosaccharides**   The simplest monosaccharide is glyceraldehyde. It has three carbon atoms (hence it is classed as a *triose*) with the formula

The central carbon atom (starred) in the molecule is asymmetric so that stereoisomers of glyceraldehyde exist. (Most naturally occurring monosaccharides are the d-isomer.)

The most important monosaccharides contain five or six carbon atoms (pentoses and hexoses respectively) and this group includes ribose, glucose and fructose.

*Ribose*   This is a pentose and it was first isolated from plant nucleic acids. (Nucleic acids play an important part in determining the sequence of amino acids in a protein.) It has the formula $CH_2OH.$-$CHOH.CHOH.CHOH.CHO$ but it can occur also in a cyclic form.

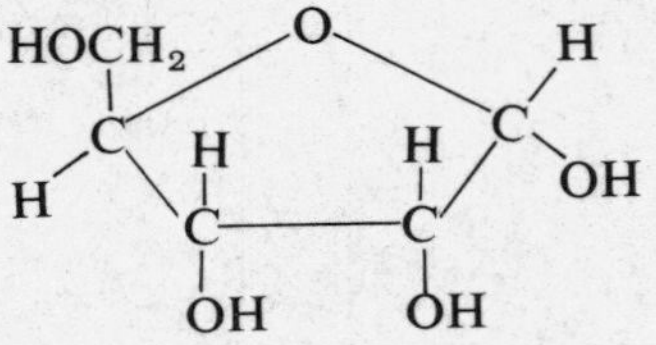

The cyclic form which contains five atoms is called a furanose ring, and results from a lone pair of electrons on the oxygen of the OH group carried by the penultimate carbon atom being attracted to the slightly positive carbon atom of the carbonyl group (see p. 72). Stereoisomers of ribose are possible according to the relative dispositions of the OH groups above and below the ring. Ribose is an important pentose as it forms part of the structure of ribonucleic acid (RNA). In a similar way, deoxyribose, which has the structure shown below, forms part of deoxyribonucleic acid, DNA.

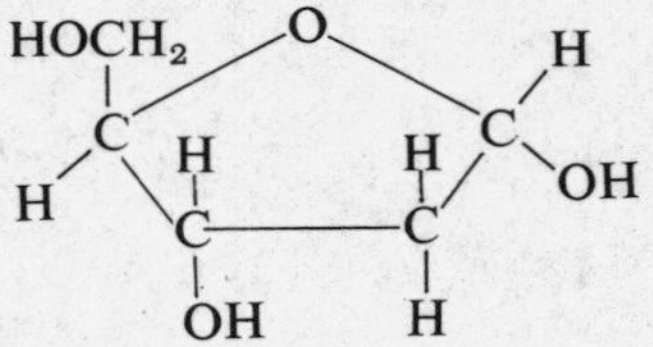

Ribose shows the typical reactions of an alcohol and an aldehyde as both these groups are present in the molecule.

*Glucose*   This is a hexose with the formula $C_6H_{12}O_6$. Again, glucose can exist as a straight chain or a cyclic molecule as shown in Figure 19.

There are eight different molecules (three of which are shown in Figure 19) corresponding to a straight-chain structure, each of which can have a d- and an l- isomer, making sixteen isomers in all. In-

144

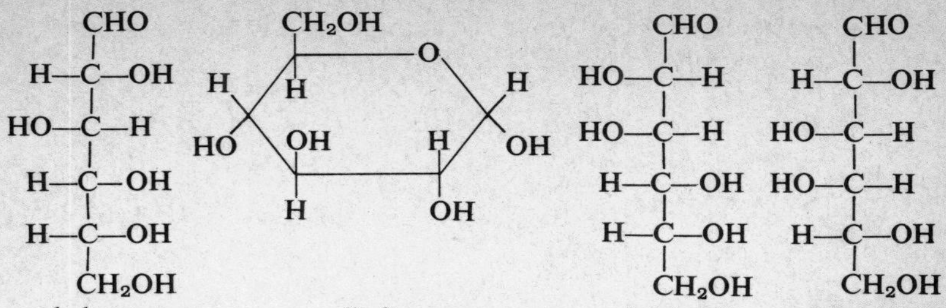

d-glucose — cyclic form of d-glucose — d-mannose — d-galactose

Fig. 19

dividual names have been assigned to the eight structures. Since these hexoses carry an aldehyde group, they are known as aldohexoses. Of the sixteen aldohexoses, a few (including those shown in Figure 19) have been isolated from natural sources, while the others have been synthesized.

*Fructose* Fructose has the dual structure

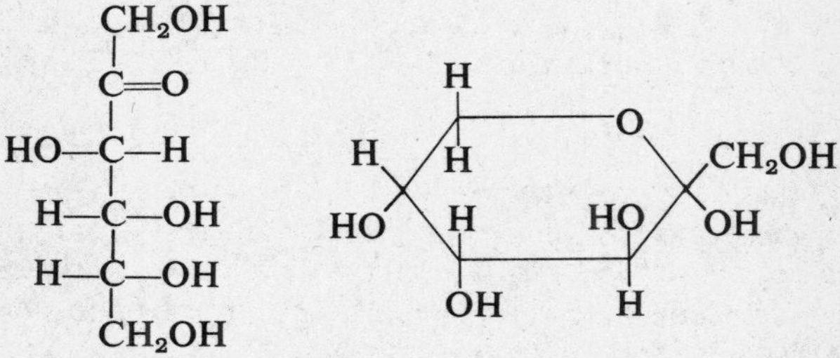

d-fructose — cyclic form of d-fructose

Since the molecule carries a ketone group, it is classed as a ketohexose, and d-fructose is the main ketohexose found in nature.

**General properties** Monosaccharides are crystalline substances, soluble in water and having a sweet taste. As distinct from disaccharides and polysaccharides, they are not hydrolysed to simpler carbohydrates. The aldehyde monosaccharides (aldoses) are good reducing agents, giving the silver mirror test and reducing Fehling's solution. Both aldo- and keto- monosaccharides undergo addition reactions, adding on hydrogen cyanide for example. They react with phenylhydrazine in the presence of ethanoic acid, to form an *osazone*, which is essentially a double phenylhydrazone.

$$
\begin{array}{l}
\text{CHO} \\
| \\
\text{CHOH} \\
| \quad + 3C_6H_5NHNH_2 \rightarrow \\
\text{(CHOH)}_3 \\
| \\
\text{CH}_2\text{OH}
\end{array}
\qquad
\begin{array}{l}
\text{H—C}\!=\!\text{NNHC}_6H_5 \\
| \\
\text{C}\!=\!\text{NNHC}_6H_5 \\
| \qquad\qquad + C_6H_5NH_2 + 2H_2O + NH_3 \\
\text{(CHOH)}_3 \\
| \\
\text{CH}_2\text{OH}
\end{array}
$$

glucose                          glucosazone

Osazones are useful derivatives for they are yellow, sparingly soluble, crystalline compounds. They usually have a definite melting point, and when viewed under a microscope, the difference between the crystalline forms of the various osazones is clear to see. These two factors make possible the identification of the different mono-saccharides, although at the present time chromatography is much more widely used for this purpose. Glucose, mannose and fructose give the same osazone which shows that the portion of the respective molecules not involved in the reaction with phenylhydrazine is the same in each case.

All four naturally occurring monosaccharides undergo fermentation by yeasts. The following mechanism refers to the fermentation of glucose, an important sugar in the metabolism of carbohydrates.

**Fermentation of glucose (glycolysis)** The breakdown or fermentation of glucose in the absence of air is known as glycolysis, and it involves a complex succession of steps referred to as the Embden–Meyerhof mechanism. Briefly, the initial stage is the phosphorylation of glucose under the influence of enzymes. Following this, fructose 1,6-diphosphoric acid is formed which breaks (along the dotted line) to give two triose (three carbon sugar) molecules, the reactions again being brought about by enzyme activity. Further reactions lead to the formation of pyruvic acid, which is decarboxylated by yeast to form ethanal and carbon dioxide.

glucose                               fructose, 1,6-diphosphoric acid

146

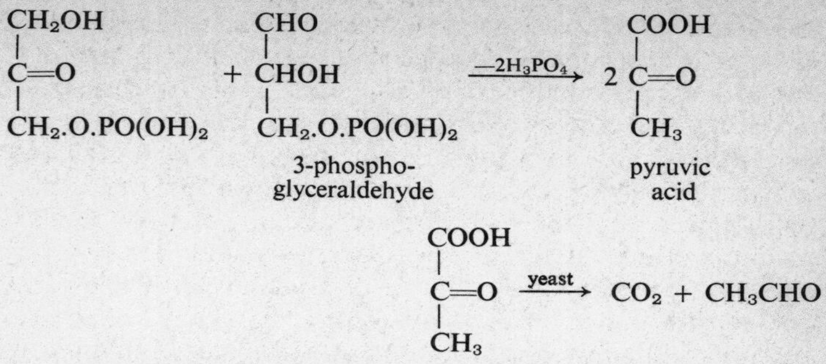

$$\begin{array}{cccc}
\text{CH}_2\text{OH} & \text{CHO} & & \text{COOH} \\
| & | & & | \\
\text{C}{=}\text{O} & + \text{CHOH} & \xrightarrow{-2\text{H}_3\text{PO}_4} & 2 \text{ C}{=}\text{O} \\
| & | & & | \\
\text{CH}_2.\text{O.PO(OH)}_2 & \text{CH}_2.\text{O.PO(OH)}_2 & & \text{CH}_3
\end{array}$$

3-phospho-<br>glyceraldehyde      pyruvic<br>acid

$$\begin{array}{c}
\text{COOH} \\
| \\
\text{C}{=}\text{O} \xrightarrow{\text{yeast}} \text{CO}_2 + \text{CH}_3\text{CHO} \\
| \\
\text{CH}_3
\end{array}$$

Fig. 20 Simplified Embden–Meyerhof mechanism

The ethanal is reduced to ethanol by reaction with the 3-phospho-glyceraldehyde formed in an earlier stage.

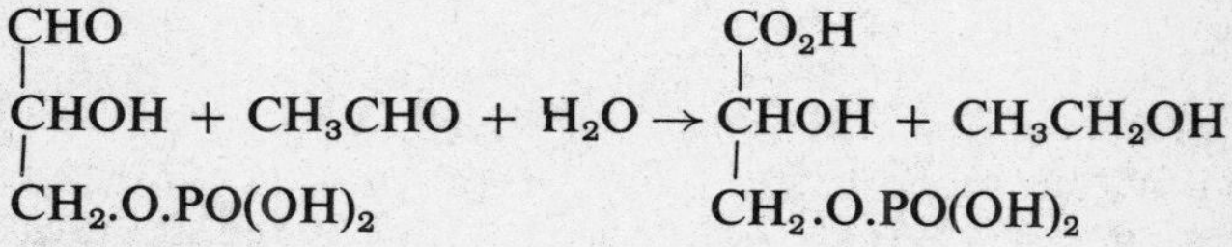

$$\begin{array}{ccc}
\text{CHO} & & \text{CO}_2\text{H} \\
| & & | \\
\text{CHOH} + \text{CH}_3\text{CHO} + \text{H}_2\text{O} \rightarrow & \text{CHOH} + \text{CH}_3\text{CH}_2\text{OH} \\
| & & | \\
\text{CH}_2.\text{O.PO(OH)}_2 & & \text{CH}_2.\text{O.PO(OH)}_2
\end{array}$$

The overall reaction is

$$C_6H_{12}O_6 \rightarrow 2C_2H_5OH + 2CO_2$$

Other enzymes, particularly those found in animals, catalyse the production of lactic acid ($CH_3CHOHCO_2H$) from pyruvic acid, so allowing for a different end product to this mechanism.

**Disaccharides** Disaccharides can be regarded as the molecules formed by the union of two monosaccharides by means of the elimination of a molecule of water. (The two monosaccharide residues need not be the same.) The most important disaccharides are sucrose (found in cane and beet sugar), maltose, and lactose (milk sugar). They are hydrolysed by enzyme action or by boiling with dilute acids.

$$C_{12}H_{22}O_{11} + H_2O \xrightarrow{H^+} C_6H_{12}O_6 + C_6H_{12}O_6$$

sucrose $\longrightarrow$ glucose + fructose
maltose $\longrightarrow$ glucose + glucose
lactose $\longrightarrow$ glucose + galactose

The hydrolysis of sucrose may be studied by measuring the change in the optical rotation of the solution using a polarimeter. Sucrose is d-rotatory, while the mixture of glucose and fructose produced by hydrolysis is l-rotatory overall. (Glucose rotates to the right, fructose rotates more strongly to the left.) Therefore, the rotation of such a solution gradually decreases as the reaction proceeds and finally changes in direction. Consequently, the reaction is referred to as *inversion* and the mixture of glucose and fructose so formed is called *invert sugar*. Sucrose is one of the chief sources of energy in our diet; it is changed into invert sugar during the initial stages of digestion by enzyme hydrolysis before being broken down further by the metabolic processes of the body. Disaccharides are sweet, soluble compounds which crystallize readily. Sucrose is not a reducing sugar—it does not reduce Fehling's solution or produce a silver mirror, neither does it form an osazone. Lactose is found in the milk of all mammals and it is rather less sweet than sucrose. It reduces Fehling's solution and forms an osazone. Maltose is made by the enzymatic hydrolysis of starch; ptyalin in saliva can bring about this change. It reduces Fehling's solution, forms an osazone and it is hydrolysed to glucose by the enzyme *maltase*.

**Polysaccharides** The repeated condensation polymerization of monosaccharide units leads to the formation of a polysaccharide molecule. Polysaccharides are usually tasteless, amorphous substances which are insoluble in water and hydrolysed by the action of enzymes or mineral acids to monosaccharides. Polysaccharides are more difficult to hydrolyse than disaccharides, and intermediate products (such as cellobiose from cellulose, and dextrins and maltose from starch) can be isolated during the course of the reaction.

The link between two monosaccharide units can be made (via the elimination of a molecule of water) in either of two ways. Numbering the carbon atoms clockwise round the ring, starting with the carbon on the right of the oxygen, two glucose units join so as to link carbon atom number 1 in the first pyranose ring to carbon atom number 4 in the second ring via an oxygen bridge. This constitutes a 1–4 link.

An α link is formed in maltose as shown above, while a β link is formed as shown below.

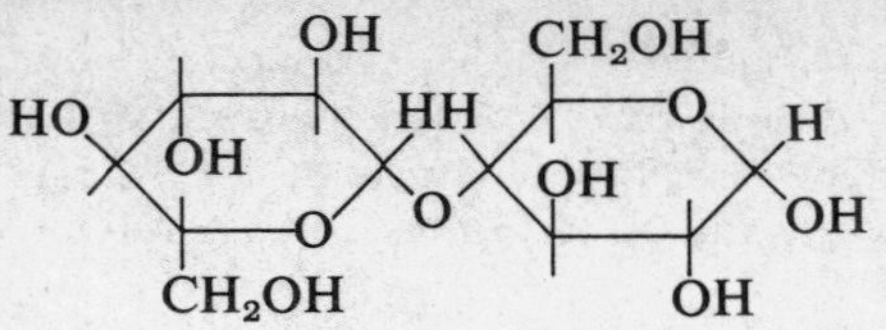

Amylose, one of the main components of starch, contains chains of up to 1000 glucose units joined by α links. Cross linking can occur which joins carbon number 1 (at the end of a chain) through an oxygen bridge to carbon number 6 (i.e. the carbon of the —CH$_2$OH group) which is already part of another chain.

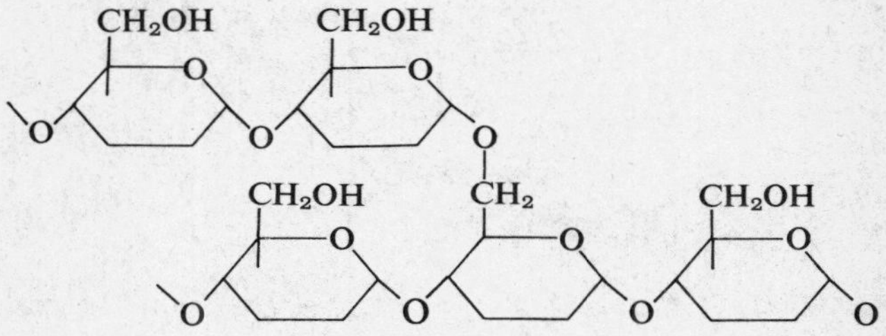

Fig. 21

This structure is present in amylopectin, the other chief constituent of starch.

Starch is a white, somewhat hygroscopic powder with no definite melting point. It does not dissolve in water, but in hot water the starch grains burst to give a milky suspension which sets on cooling to give starch paste. Starch gives an intense blue colour with iodine.

Cellulose consists of glucose units joined by β linkages to form long unbranched chains. These chains pack together to form a strong fibrous tissue. It is interesting to note that an enzyme hydrolyses either an α or a β linkage, but not both. Thus, the human body, which does not possess an enzyme capable of attacking a β link, is unable to digest cellulose.

Cellulose is soluble in concentrated sulphuric acid; on diluting and boiling the solution for some time, hydrolysis to glucose is effected. Cellulose can be nitrated (using a mixture of concentrated sulphuric and nitric acids) to form cellulose nitrate which is used as an

149

explosive. Treatment of cellulose with ethanoic anhydride produces cellulose acetate (ethanoate) which is soluble in propanone. When a fine jet of such a solution is sprayed into a bath of dilute acid or alkali, a thread of artificial silk is formed. A similar process takes place when cellulose is treated with sodium hydroxide followed by carbon disulphide—the product (cellulose xanthate) is called viscose. Viscose and acetate rayons are important man-made fibres.

## Fats

The group of naturally occurring esters of fatty acids together with other oil-soluble derivatives of fatty acids are collectively classed as lipids. Fats are esters of glycerol (propane 1,2,3-triol), and three sets of esters (called glycerides) can be produced (mono-, di- and tri-glycerides) according to the number of OH groups esterified. Natural fats and oils consist of triglycerides only, for example, tristearin or glyceryl tristearate is

$$\begin{array}{l} CH_2OCOC_{17}H_{35} \\ | \\ CHOCOC_{17}H_{35} \\ | \\ CH_2OCOC_{17}H_{35} \end{array}$$

The acid residues found in fats are derived from the following acids:

| | | | |
|---|---|---|---|
| lauric acid | $C_{11}H_{23}CO_2H$ | m.p. 44°C | (saturated) |
| myristic acid | $C_{13}H_{27}CO_2H$ | m.p. 58° | (saturated) |
| palmitic acid | $C_{15}H_{31}CO_2H$ | m.p. 63° | (saturated) |
| stearic acid | $C_{17}H_{35}CO_2H$ | m.p. 72° | (saturated) |
| oleic acid | $C_{17}H_{33}CO_2H$ | m.p. 15° | (unsaturated) |
| linoleic acid | $C_{17}H_{31}CO_2H$ | m.p. −11° | (unsaturated) |

The differences between the various fats depend entirely on the differences between the acid radicals present in the triglyceride. Two important points may be noted:

(a) As the degree of unsaturation among the acids increases, the melting point of the acid, and hence the fat, decreases. For example, the relative molecular masses of linoleic, oleic and stearic acids are 280, 282 and 284 respectively, but their melting points are −11°, 15° and 72°C. Thus, a fat containing a high proportion of stearate radicals has a high melting point, while one containing a large linoleate content is likely to be an oil at room temperatures.

(b) A double bond is a reactive centre in a molecule. It is possible
for the double bond in the acid residue in an unsaturated fat
to break under the influence of heat, moisture and air, liberating acids of a shorter chain length as a result:

$$
\begin{array}{c}
\text{H} \quad \text{H} \quad \text{H} \quad \text{H} \\
| \quad\;\; | \quad\;\; | \quad\;\; | \\
-\text{C}-\text{C}=\text{C}-\text{C}- \quad \xrightarrow[\text{O}_2]{\text{H}_2\text{O}} \quad
\begin{array}{c}\text{H}\\ |\\ -\text{C}-\text{C}\end{array}\!\!\!\begin{array}{c}\text{OH}\\ \\ \text{O}\end{array}
\;+\;
\begin{array}{c}\text{HO}\\ \\ \text{O}\end{array}\!\!\!\begin{array}{c}\text{H}\\ |\\ \text{C}-\text{C}-\\ |\\ \text{H}\end{array} \\
| \quad\;\; | \\
\text{H} \quad\;\; \text{H}
\end{array}
$$

These short-chain acids often have unpleasant odours and
tastes (e.g. butanoic acid has a strong cheese-like odour) and
the liberation of such acids gives rise to rancidity in fats.

Fats are colourless, odourless and tasteless when pure, but natural
fats contain other materials which lend colour, taste and odour to the
material. Because they are complex mixtures, fats have no definite
melting points although, in general, increasing unsaturation in the
fat lowers the melting range.

**Hardening of fats and oils**   In the presence of a nickel catalyst, an
unsaturated oil or fat will add on hydrogen. Hydrogenation is usually
carried out at temperatures between 140° and 180°C with the catalyst
suspended in the liquid. Purified oils may be converted into clean
white products having higher melting points, and in view of the
increase in melting point brought about, this reaction is referred to
as the *hardening* of fats and oils. Hardened oils from sources such as
palm kernel, ground nut, soya-bean and whale oils when blended
with other constituents form margarine, which is an important food-
stuff.

Refluxing a fat with alkali produces glycerol and soap, the hydrolysis being termed *saponification*.

**Soap and detergents**   A soap molecule, sodium stearate,
$C_{17}H_{35}COO^-Na^+$, for example, consists of a long hydrocarbon

$$
\text{H}-\underset{\underset{\text{H}}{|}}{\overset{\overset{\text{H}}{|}}{\text{C}}}-\underset{\underset{\text{H}}{|}}{\overset{\overset{\text{H}}{|}}{\text{C}}}-\underset{\underset{\text{H}}{|}}{\overset{\overset{\text{H}}{|}}{\text{C}}}-\underset{\underset{\text{H}}{|}}{\overset{\overset{\text{H}}{|}}{\text{C}}}-\underset{\underset{\text{H}}{|}}{\overset{\overset{\text{H}}{|}}{\text{C}}}-\underset{\underset{\text{H}}{|}}{\overset{\overset{\text{H}}{|}}{\text{C}}}-\underset{\underset{\text{H}}{|}}{\overset{\overset{\text{H}}{|}}{\text{C}}}-\underset{\underset{\text{H}}{|}}{\overset{\overset{\text{H}}{|}}{\text{C}}}-\underset{\underset{\text{H}}{|}}{\overset{\overset{\text{H}}{|}}{\text{C}}}-\underset{\underset{\text{H}}{|}}{\overset{\overset{\text{H}}{|}}{\text{C}}}-\underset{\underset{\text{H}}{|}}{\overset{\overset{\text{H}}{|}}{\text{C}}}-\underset{\underset{\text{H}}{|}}{\overset{\overset{\text{H}}{|}}{\text{C}}}-\underset{\underset{\text{H}}{|}}{\overset{\overset{\text{H}}{|}}{\text{C}}}-\underset{\underset{\text{H}}{|}}{\overset{\overset{\text{H}}{|}}{\text{C}}}-\underset{\underset{\text{H}}{|}}{\overset{\overset{\text{H}}{|}}{\text{C}}}-\underset{\underset{\text{H}}{|}}{\overset{\overset{\text{H}}{|}}{\text{C}}}-\underset{\underset{\text{H}}{|}}{\overset{\overset{\text{H}}{|}}{\text{C}}}-\text{C}\!\!\begin{array}{c}\nearrow\text{O}^-\\ \searrow\text{O}\end{array}\;\;\text{Na}^+
$$

'tail' and a carboxylate ion 'head'. The hydrocarbon tail is soluble
in grease, while the carboxylate head dissolves in water. A globule
of grease is emulsified in water by the action of many soap molecules,
as shown in Figure 22. The soap molecules attach themselves by the

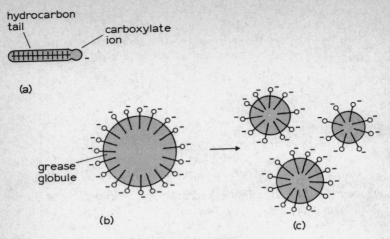

Fig. 22

hydrocarbon tails to the globules of grease; thus a large concentration of negatively charged carboxylate ions protrude through the surface of the globules into the water. Provided there are sufficient soap molecules present, the globules are repeatedly broken down to form an emulsion. This action is known as detergency; both soap and synthetic detergents (long-chain hydrocarbons carrying a water-soluble end) function in this way. The presence of calcium and magnesium salts in hard water precipitates the soap molecules as scum and prevents the detergent action.

## Amino Acids and Proteins

Amino acids form the basic units from which protein molecules are built up. About twenty amino acids have been isolated from the hydrolysis of proteins and in this field chromatographic techniques of identification and quantitative estimation have proved invaluable.

The general formula for an amino acid is $\begin{array}{c} R \\ \diagdown \\ H_2N \end{array} CHCO_2H$ where R is a side chain which may be a simple alkyl group, or something much more complex. Some amino acids are shown in Table 23.

Aminoethanoic acid, or glycine, is the simplest amino acid. It has the formula $H_2NCH_2CO_2H$, and in aqueous solution it exists as a dipolar ion, or zwitterion $H_3^+NCH_2COO^-$, where the proton from the carboxyl group has moved to the basic $NH_2$ group. All naturally

152

TABLE 23 *Some amino acids*

| Identity of R | Amino acid | Abbreviation |
|---|---|---|
| —H | glycine | gly |
| —$CH_3$ | alanine | ala |
| —$CH_2OH$ | serine | ser |
| —$CH_2SH$ | cysteine | cys |
| —$CH_2CH_2OH$ | threonine | thr |
| —$CH_2C_6H_5$ | phenylalanine | phe |
| —$CH_2C_6H_4OH$ | tyrosine | tyr |
| —$CH_2COOH$ | aspartic acid | asp |
| —$(CH_2)_4NH_2$ | lysine | lys |

occurring amino acids apart from glycine, are optically active in the 'l' sense.

Pure amino acids are white solids, usually soluble in water giving a neutral solution, as the effect of the carboxylic group is compensated by the basic amino group. However, if the side chain R contains an acid group, the aqueous solution is acid, while it is alkaline if R contains an amino group.

Amino acids can link together through peptide links, formed by the elimination of water between the carboxyl and amino groups.

$$—CO_2H + H_2N— \rightarrow H_2O + —CONH—$$

When a number of amino acids are combined in this way, a polypeptide molecule is formed; larger structures involving fifty or more amino-acid residues are called proteins. Proteins, in spite of being made up from a large number of units, are not bulky, sprawling molecules, rather they are compact, close-knit structures. The properties of a protein are largely governed by the sequence in which the various amino acids appear in the chain and by the way they (and in particular, their side chains) interact and dovetail with each other.

**The Chemical analysis of proteins** The task of discovering the identity and sequence of the amino acids present in a protein molecule is formidable. However, in 1954, as a result of many years of investigation using established techniques and after developing new methods of analysis, the amino-acid sequence in insulin was finally determined. At the start of the investigation, insulin was the only pure protein available. On refluxing for twenty-four hours with 6M hydrochloric acid, insulin is completely broken down into free amino acids. By means of chromatographic techniques, it was found that fifty-one amino-acid molecules were combined in the insulin molecule and that seventeen different amino acids were present (e.g. six

153

cysteine, one lysine etc.). The next step in the analysis was to determine the sequence in which these units were joined together. This stage is called N-terminal analysis. The aromatic compound 1-fluoro-2,4-dinitrobenzene (or FDNB) has the ability to react with the free amino group at the end of a peptide chain. The stages are shown in Figure 23.

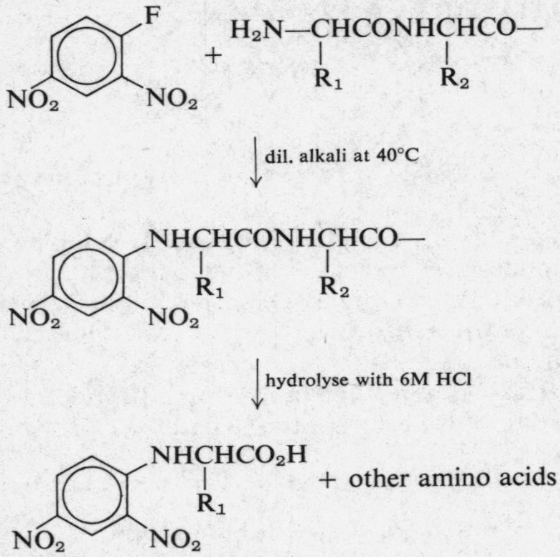

Fig. 23

When the product formed between the peptide chain and FDNB is hydrolysed with 6M hydrochloric acid, the terminal amino-acid residue is retained by the aromatic molecule. The identity of this product can be determined chromatographically. Further progress is made along similar lines using less drastic methods of hydrolysis, so that the protein does not break down into separate amino acids, but polypeptides of moderate chain length are produced. Hydrolysis by enzymes is very important in this respect as they choose to attack only a particular type of bond; for example, the enzyme chymotrypsin usually breaks peptide links formed between amino acids having aromatic side chains.

The determination of the number and sequence of amino-acid residues in a protein molecule marks the end of the first stage in the elucidation of the structure of a protein. This is called the *primary structure*. X-ray diffraction studies of proteins reveal the disposition of the atoms and groups in space. This three-dimensional picture of

154

a protein molecule is called the *secondary structure*. Two characteristic secondary features are apparent:

(a) $\beta$-pleated sheets
(b) $\alpha$-helix.

These regular structures occur because the regular disposition of —CO— and —NH— groups along the peptide chain allows for the formation of hydrogen bonds. In proteins, the $\alpha$-helix and the $\beta$-pleated sheets are the struts and plates which reinforce the protein structure. Another important stabilizing influence is the formation of disulphide bridges resulting from the oxidation of two cysteine (pronounced cys-tay-een) residues to give cystine.

$$
\begin{array}{c}
| \\
NH \\
| \\
CH{-}CH_2{-}SH \\
| \\
CO \\
|
\end{array}
+
\begin{array}{c}
| \\
NH \\
| \\
HS{-}CH_2{-}CH \\
| \\
CO \\
|
\end{array}
\longrightarrow
$$

$$
\begin{array}{c}
| \\
NH \\
| \\
CH{-}CH_2{-}S{-}S{-}CH_2{-}CH \\
| \\
CO \\
|
\end{array}
\qquad
\begin{array}{c}
| \\
NH \\
| \\
\\
| \\
CO \\
|
\end{array}
$$

The secondary structure of a protein is influenced by the identity of the side chains carried by the amino-acid residues, especially in relation to their size and the effect of a change in pH on them. Moderate heating (up to 75°C), strong acids or alkalies or any reagent which causes the hydrogen bonding to weaken or break, disrupts the secondary structure. If this happens, the activity of the protein is lost and it is said to be *denatured*. A denatured protein cannot regain its secondary structure. Finally, the plate and strut secondary structure is further convoluted to form an even more compact structure called the *tertiary structure* (somewhat analogous to a coiled-coil electric light filament). Once again, the tertiary structure is held in place by the interaction (especially hydrogen bonding and disulphide bridges) between the side chains of the amino-acid residues.

**Enzymes**   Enzymes are a particular class of proteins whose function is to catalyse biochemical reactions. The substance changed by enzyme action is called the *substrate* and enzymes are often named according to the substrate they affect. For example, maltase hydrolyses maltose to glucose, while urease converts urea into ammonia, also by hydrolysis. Some of the most important characteristics of enzymes are:

(a) *High catalytic efficiency*. The catalytic power of an enzyme usually greatly exceeds that of a non-biological catalyst. As an example, one molecule of an enzyme can bring about the decomposition of many millions of molecules of substrate per minute.

(b) *Specificity of action*. Very many enzymes will attack only one optical isomer, or a certain group or particular kind of bond. Some enzymes, for example, will hydrolyse peptide links which join amino acids having aromatic side chains, while others attack a wider range of peptide links.

(c) *Influence of pH*. Enzymes function best at a certain pH (the optimum pH) generally in the range 4·5 to 7.

(d) *Influence of temperature*. Again, an optimum value exists, at which temperature the enzyme functions most effectively. Lowering or increasing the temperature from this optimum value reduces enzyme efficiency. Because proteins are denatured on heating, enzyme action is brought to a standstill when the temperature is taken above a certain upper limit.

In view of their efficiency and specificity, enzymes are widely used in industry. Typical examples of their use range from fermentation in the production of beer, wines and spirits and in bread making to their incorporation as additives in foodstuffs, washing powders and pharmaceutical products.

### Nucleic Acids

Nucleic acids are normal cell constituents of all living material, and their function is to control the production of the various proteins required by the living system. There are two distinct types of nucleic acids, DNA (deoxyribose nucleic acid) and RNA (ribose nucleic acid). Ribose nucleic acid molecules exist as single separate chains which are made up of ribose residues linked by phosphate bridging

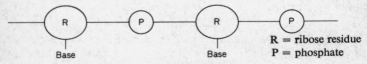

R = ribose residue
P = phosphate

Fig. 24 RNA structure

156

groups. Each sugar molecule is further linked to an organic base as shown diagrammatically in Figure 24.

Five organic bases are found combined in nucleic acids. Cytosine, guanine and adenine are present in both RNA and DNA, while uracil occurs only in RNA and thymine is confined to DNA. The sequence in which the bases appear in the chain is critical as it forms the basis for the 'genetic code' which directs the building up of proteins by cells and the self-reproduction of cells.

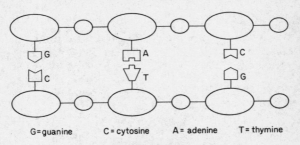

G = guanine    C = cytosine    A = adenine    T = thymine

Fig. 25  Diagram of DNA structure

The DNA structure was shown by Crick and Watson in the 1950s to consist of two intertwined chains. The two strands are knit together by means of repeated hydrogen bonding between a pair of bases (one on each chain). On account of their molecular size and shape, cytosine will pair up only with guanine, while adenine pairs only with thymine. A simplified diagram of the DNA structure is shown in Figure 25. As with proteins, DNA has a secondary structure as the primary twin strand is twisted to form a double helix.

# 18   Industrial Organic Chemicals

The main sources of organic chemicals are coal, petroleum and natural gas.

## Coal

Heating coal in the absence of air yields four main products: coal gas, ammoniacal liquor, coal tar and coke. The chief chemicals produced from these sources are shown in Figure 26.

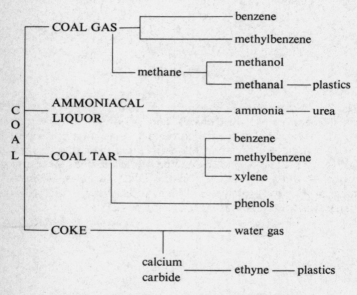

Fig. 26 Chief organic chemicals produced from coal

Coal tar contains a wide variety of aromatic compounds, while methane and ethyne may be recovered from coal gas. Benzene and methylbenzene are recovered from coal gas (p. 97) and from coal tar, and at present coal forms the major source of both. Coke oven gas contains ethene, which may be converted into ethanal, polyethene and ethylbenzene. Dehydrogenation of ethylbenzene produces styrene, which polymerizes to form an important plastic, polystyrene.

**Petroleum**

Apart from the hydrocarbons used as fuel for internal combustion and jet engines, the petroleum industry manufactures over two million tons of organic chemicals (in the UK) each year. In addition, over half a million tons of inorganic chemicals (mainly carbon black, ammonia, hydrogen cyanide and sulphuric acid) are produced from petroleum sources. Figure 27 gives an indication of the industrial demand for petrochemicals.

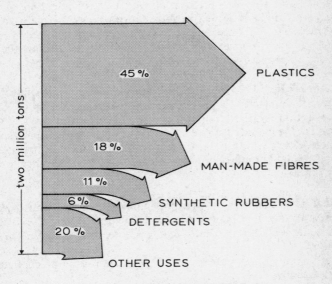

Fig. 27 Major outlets of petrochemicals

Petroleum consists of aliphatic hydrocarbons (mainly alkanes), some aromatic hydrocarbons (the content varies according to the source of the petroleum crudes), and some sulphur-, oxygen- and nitrogen-containing compounds.

**Petroleum refining** This is chiefly concerned with the production of fuel for internal combustion, including jet engines. Distillation of a petroleum crude yields:

    (a) gas—mainly methane and ethane,
    (b) liquefiable petroleum gas—this liquefies under pressure and is sold as a fuel in the form of 'bottled' gas,
    (c) petroleum ether—mainly pentanes and hexanes,

159

(d) light naphtha—mainly hexanes and heptanes,
(e) petrol,
(f) paraffin (kerosene),
(g) gas oil,
(h) lubricating oil.

The successive fractions show a steady increase in boiling point. Only a small proportion of the output—petrol and paraffin—is suitable as engine fuel. The various processes of catalytic cracking, hydroforming and aromatization are aimed at producing fuel of good quality from higher boiling fractions. During cracking (breakdown of large molecules) quantities of carbon and lower hydrocarbons are produced. The hydrocarbons, together with the petroleum ether and light naphtha fractions, are passed to the petrochemical plant.

**Petrochemicals** The feed to the petrochemical plant is subjected to further cracking, and the chief products are shown in Figure 28.

In addition, synthesis gas (hydrogen and carbon monoxide) which is used for the manufacture of ammonia and methanol is produced by the petrochemical industry. Coal remains the major source of benzene, although an increasing quantity is being produced from petroleum sources.

**Natural gas** This contains carbon dioxide, nitrogen and lower alkanes, with methane as the major constituent. At present, North Sea gas is being used as a fuel, although it is potentially a useful source of chemicals.

The manufacture of a selection of important organic chemicals is outlined below.

**Detergents**

Sodium alkyl sulphonates are frequently used as detergents. They are made by reacting straight-chain alkanes with sulphur dioxide and chlorine in ultraviolet light.

$$C_{14}H_{30} + SO_2 + Cl_2 \rightarrow C_{14}H_{29}SO_2Cl + HCl$$

(Alkanes containing between 12 and 16 carbon atoms are the most suitable.) The product is treated with sodium hydroxide solution.

$$C_{14}H_{29}SO_2Cl + 2NaOH \rightarrow NaCl + H_2O + C_{14}H_{29}SO_3{}^-Na^+$$

When branched-chain hydrocarbons are used for this process, the detergents produced are not easily broken down by micro-organisms and therefore cause problems related to their disposal. Straight-chain

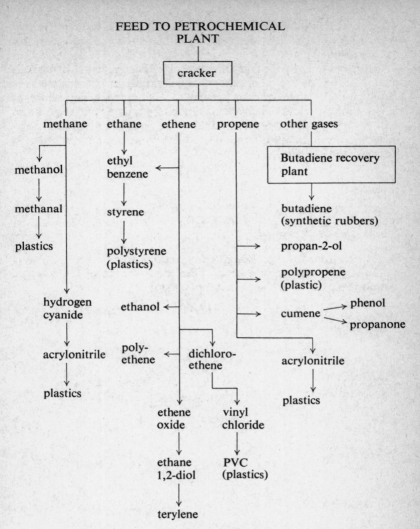

## FEED TO PETROCHEMICAL PLANT

cracker

methane · ethane · ethene · propene · other gases

methanol

methanal

plastics

hydrogen cyanide

acrylonitrile

plastics

ethyl benzene

styrene

polystyrene (plastics)

ethanol

poly-ethene

Butadiene recovery plant

butadiene (synthetic rubbers)

propan-2-ol

polypropene (plastic)

cumene → phenol

cumene → propanone

acrylonitrile

plastics

dichloro-ethene

ethene oxide

ethane 1,2-diol

terylene

vinyl chloride

PVC (plastics)

Fig. 28  Chemicals from petroleum

hydrocarbons give rise to soft or biodegradable detergents. The problem of isolating straight-chain hydrocarbons from a mixture was solved by the use of molecular sieves. These are special silicate structures (zeolites) which have been processed to remove the water normally retained within the structure. The molecular sieves so formed have spaces, or pores, within the crystal lattice wide enough to allow the passage of molecules having a small diameter (e.g. straight-chain hydrocarbons).

Alkyl benzene sulphonates are also used as detergents. Benzene is made to react with a chlorinated alkane (e.g. $C_{11}H_{23}Cl$, the chlorine atom must be carried by the terminal carbon) in the presence of aluminium chloride. The product is then sulphonated and the sodium salt (which is the detergent) is obtained from this. Fluorescent powders and enzymes are often added to detergent powders in order to increase their general effect.

## Plastics

Plastics are defined as compounds which at some stage in their manufacture are present in a readily deformable state. Plastics may be produced by the following types of reactions:

1 *Polymerization due to breakages of multiple bonds.* An example is the polymerization of ethene under the influence of temperature and pressure, or in the presence of a catalyst:

$$nCH_2{=}CH_2 \rightarrow (-CH_2-CH_2-)_n$$

Polystyrene is produced by way of the following steps:

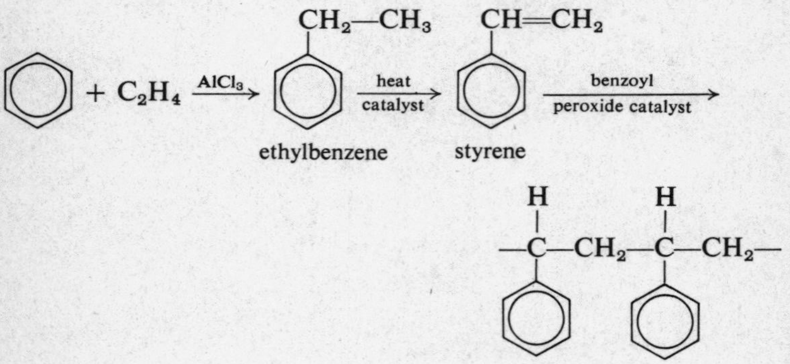

ethylbenzene  styrene

polystyrene

Polyvinyl compounds are produced by this type of reaction also:

$$CH_2{=}CH_2 + Cl_2 \rightarrow CH_2ClCH_2Cl \xrightarrow{heat} CH_2{=}CHCl + HCl$$

Then,

$$nCH_2{=}CHCl \rightarrow (-CH_2.CHCl-)_n$$
PVC

**2** *Condensation polymerization.* The basic units forming the plastic are linked following the elimination of the elements of water between two molecules (condensation reaction). The polyamide condensation is an example:

$$H_2NCH_2COOH + H_2NCH_2COOH \rightarrow$$
$$-NH-CH_2-CO-NH-CH_2-CO- + H_2O$$

Alternatively, both amino groups could be located on one molecule and both acid groups on the other. For example,

$$H_2N(CH_2)_6NH_2 + HOOC(CH_2)_4COOH$$

hexamethylene        adipic acid
diamine

$$\xrightarrow{-H_2O} -NH(CH_2)_6NH-CO-(CH_2)_4-CO-\text{etc.}$$

polyamide plastic (nylon 6,6)

In condensation plastics, two usually different molecules are joined together in 'head to tail' fashion, and this allows several variations to be incorporated in the same type of plastic. Thus a range of nylon polymers has been produced.

Another type of condensation polymer results from the elimination of water between a carboxyl and a hydroxyl group (esterification), and such polymers are known as *polyesters*. For example,

terephthalic acid      ethane-1,2-diol

terylene

The polymers resulting from these processes are termed *resins*. To convert them into a useful plastic, a mixture of resins is blended, together with softening agents (plasticizers) and antioxidants.

Two types of plastic are recognized: thermoplastic materials which soften again on heating, and thermosetting plastics which do not.

A great advance in plastics technology was made following the introduction of the Ziegler catalysts. Instead of polymerizing ethene at 1500 atmospheres and 200°C, the Ziegler process uses a temperature of about 70°C and the pressure is 6 atmospheres. Ethene is bubbled into a hydrocarbon solvent in which the catalyst (a mixture of titanium tetrachloride and an alkyl derivative of aluminium) is suspended, and the resultant polyethene is found to have a greater density, higher softening point and greater tensile strength. The reason for this is that catalytic polymerization joins the monomer molecules together in an exact order, giving the final polymer a regular shape as opposed to the haphazard linking produced by the high-pressure technique. Polypropene has been produced by this method, and the product is a stereoregular (i.e. exactly ordered in space) polymer known as isotactic polypropene. This is a very important polymer; it is extremely tough, it has a high tensile strength and the lowest density ($0.906 \text{ g cm}^{-3}$) of all commercial plastics. Modifications of the catalytic polymerization technique have led to the production of synthetic rubbers with properties at least equal and often better than natural rubber.

**Butadiene** ($CH_2{=}CH{-}CH{=}CH_2$)   Apart from sucrose, butadiene is the organic chemical produced in the greatest quantity on a world basis. It is the starting point for the production (by polymerization, and co-polymerization with other unsaturated molecules) of a vast range of rubbers and rubber-like materials. It is produced by the vapour phase dehydrogenation of butene, using a catalyst which is usually based on iron, nickel and chromium oxides.

**Ethene oxide**   This compound, made by the controlled oxidation of ethene, is largely used as an intermediate in the manufacture of ethane-1,2-diol (ethylene glycol), which is used as an antifreeze and for the production of non-ionic detergents and a range of other compounds.

$$2CH_2{=}CH_2 + O_2 \xrightarrow[\text{catalyst}]{\text{Ag}} 2CH_2{-}CH_2$$
$$\overset{\diagdown}{\phantom{x}}O\overset{\diagup}{\phantom{x}}$$

**Dimethylamine**   The methylamines are made by the reaction between ammonia and methanol.

$$CH_3OH + NH_3 \rightarrow CH_3NH_2 + H_2O$$

$$CH_3NH_2 + CH_3OH \rightarrow H_2O + CH_3NHCH_3 \quad \text{etc.}$$

An aluminium oxide catalyst is used at a temperature of 400°C and a pressure of 60 atmospheres. Dimethylamine is made in the largest quantity as it is needed for the production of solvents for man-made fibres, and in the manufacture of jet and rocket fuels, detergents and agricultural chemicals.

**Nitrobenzene and phenylamine**   On a large scale, benzene is nitrated using nitrating acids (concentrated nitric and sulphuric acids) at

164

50 to 55°C. The mixture is agitated thoroughly and at the end of the reaction, the nitrobenzene is separated from the acid and washed with water. Most of the product is reduced to phenylamine (aniline), usually by means of iron turnings and dilute hydrochloric acid. The phenylamine is recovered by steam distillation following the neutralization of the reaction mixture with calcium hydroxide. Although phenylamine is important in the manufacture of dyes, more than half the output is used in the rubber industry.

# 19  Summary of Reactions

## Grignard Reagents

A Grignard reagent is formed by the reaction of a halogeno-alkane and magnesium in dry ethoxy-ethane (ether):

$$R.X + Mg \rightarrow R.MgX$$

Some reactions of Grignard reagents are as follows:

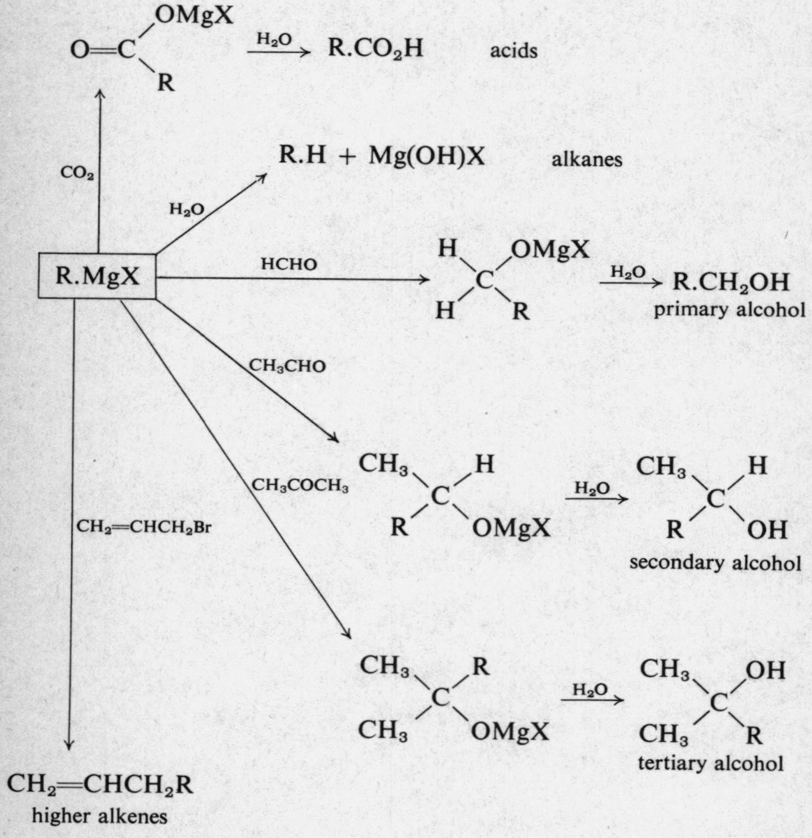

$$O = C \begin{array}{c} OMgX \\ \\ R \end{array} \xrightarrow{H_2O} R.CO_2H \qquad acids$$

$$R.H + Mg(OH)X \qquad alkanes$$

R.MgX

HCHO:

$$H - C \begin{array}{c} OMgX \\ \\ R \end{array} \xrightarrow{H_2O} R.CH_2OH$$
primary alcohol

$CO_2$

$H_2O$

$CH_3CHO$:

$$CH_3 - C \begin{array}{c} H \\ \\ OMgX \end{array} \xrightarrow{H_2O} CH_3 - C \begin{array}{c} H \\ \\ OH \end{array}$$
secondary alcohol

$CH_3COCH_3$:

$$CH_3 - C \begin{array}{c} R \\ \\ OMgX \end{array} \xrightarrow{H_2O} CH_3 - C \begin{array}{c} OH \\ \\ R \end{array}$$
tertiary alcohol

$CH_2{=}CHCH_2Br$

$CH_2{=}CHCH_2R$
higher alkenes

## Ascent of Homologous Series

### From hydrocarbons

$$CH_4 \xrightarrow{Br_2} CH_3Br \xrightarrow{2Na} CH_3CH_3$$

$$\downarrow KCN$$

$$CH_3CN \xrightarrow{H_2O} CH_3CO_2H$$

$$\downarrow \text{reduce}$$

$$CH_3CH_2NH_2 \xrightarrow{HNO_2} CH_3CH_2OH$$

### From aldehydes and alcohols

$$HCHO \xrightarrow{H_2} CH_3OH \xrightarrow{PBr_3} CH_3Br \xrightarrow{2Na} CH_3CH_3$$

$$\downarrow KCN$$

$$CH_3CN \xrightarrow{H_2O} CH_3CO_2H$$

$$\downarrow \text{reduce}$$

$$CH_3CH_2NH_2 \xrightarrow{HNO_2} CH_3CH_2OH$$

## Descent of Homologous Series

$$CH_3CH_3 \xrightarrow{Cl_2} CH_3CH_2Cl \xrightarrow{NaOH} CH_3CH_2OH \xrightarrow{(O)} CH_3CHO$$

$$\downarrow (O)$$

$$CH_3CO_2H$$

$$CH_3COCl \xleftarrow{PCl_3} \qquad \downarrow NaOH$$

$$CH_3CO_2Na$$

$$\downarrow NH_3 \qquad \downarrow \text{heat} + NaOH$$

$$CH_3CONH_2 \qquad\qquad CH_4$$

$$\downarrow NaOBr$$

$$CH_3NH_2 \xrightarrow{HNO_2} CH_3OH \text{ (in very low yield)}$$

## Types of Reaction

**Substitution reactions**   One group or atom in a compound is replaced by another.

$$CH_4 + Cl_2 \rightarrow HCl + CH_3Cl$$

167

**Addition reactions** Two molecules are linked due to the opening of a multiple bond.

$$CH_2{=}CH_2 + Br_2 \rightarrow CH_2BrCH_2Br$$

**Condensation reactions** Two molecules link following the elimination of a molecule of water (or other similar small molecule) between them.

$$CH_3CHO + H_2NOH \rightarrow H_2O + CH_3CH{=}NOH$$

**Esterification** Reaction between an alcohol and an acid through the elimination of a molecule of water.

$$R.CO_2H + R'.OH \rightleftharpoons R.COOR' + H_2O$$

**Hydrolysis** The formation of a product by addition of the elements of water to a reactant (often effected by alkali or acid).

$$CH_3CONH_2 + H_2O \xrightarrow{NaOH} CH_3CO_2H + NH_3$$

**Polymerization** The linking of many small molecules to form a large structure consisting of many repeating units.

(a) Addition polymerization:

$$nCH_2{=}CH_2 \rightarrow (-CH_2CH_2-)_n$$

(b) Condensation polymerization:

$$HCHO + H_2N-\underset{\underset{O}{\|}}{C}-NH_2 \xrightarrow{-H_2O} \left( -\underset{\underset{H}{|}}{\overset{\overset{H}{|}}{C}}-NHC\underset{\underset{O}{\|}}{\phantom{|}}NH- \right)_n$$

**Ethanoylation and benzoylation** Introduction of the $CH_3CO-$ or the $C_6H_5CO-$ groups respectively into a molecule. Usually brought about by reaction with ethanoyl (acetyl) chloride or benzoyl chloride.

$$R.OH + CH_3COCl \rightarrow HCl + CH_3COOR$$

$$R.NH_2 + C_6H_5COCl \rightarrow HCl + C_6H_5CONHR$$

**Ammonolysis** The production of an alcohol from an ester by reaction with ammonia.

$$CH_3COOC_2H_5 + NH_3 \rightarrow CH_3CONH_2 + C_2H_5OH$$

**Friedel–Crafts reaction**   Preparation of higher homologues using an aluminium chloride catalyst.

$$C_6H_6 + CH_3Cl \xrightarrow{AlCl_3} C_6H_5CH_3 + HCl$$

$$C_6H_6 + CH_3COCl \xrightarrow{AlCl_3} C_6H_5COCH_3 + HCl$$

**Wurtz reaction; Fittig reaction**   Preparation of higher hydrocarbons from halogenated hydrocarbons.

*Wurtz:*   $2CH_3Br + 2Na \rightarrow 2NaBr + CH_3.CH_3$

*Wurtz–Fittig:*   $2C_6H_5Br + 2Na \rightarrow 2NaBr + C_6H_5.C_6H_5$

**Hofmann's degradation**   Formation of primary amines from amides.

$$CH_3CONH_2 + Br_2 + 4KOH \rightarrow$$

$$CH_3NH_2 + 2KBr + K_2CO_3 + 2H_2O$$

**Rosenmund reaction**   Production of aldehydes by reduction of acid chlorides by hydrogen in the presence of a partially poisoned palladium catalyst.

$$R.COCl + H_2 \rightarrow HCl + R.CHO$$

**Cannizzaro reaction**   One molecule of an aldehyde reduces another.

$$2HCHO \xrightarrow{NaOH} HCO_2H + CH_3OH$$

**Decarboxylation**   The loss of a carboxyl group on heating the sodium salt of an acid with soda lime.

$$R.CO_2Na + NaOH \rightarrow R.H + Na_2CO_3$$

# Bibliography

ABBOTT, D. and ANDREWS, R. S. 1965. *Introduction to chromatography*. London: Longman.

BADDELEY, G. 1964. *A fragment of stereochemistry*. Education in Chemistry 1. London: The Chemical Society.

BUTTLE, J. W. and DANIELS, D. J. 1967. *Practical chemistry—an integrated course*. London: Butterworth.

FIESER, L. F. and FIESER, M. 1966. *Introduction to organic chemistry*. Farnborough: D. C. Heath.

FINAR, L. 1973. *Organic chemistry*. London: Longman.

MACKENZIE, C. A. 1962. *Unified organic chemistry*. New York: Harper & Row.

SAMUEL, D. M. 1972. *Industrial chemistry—organic*, 2nd edn. London: The Chemical Society.

SYKES, P. 1970. *A guidebook to mechanism in organic chemistry*, 3rd edn. London: Longman.

WADDAMS, A. L. 1973. *Chemicals from petroleum*, 3rd edn. London: John Murray.

172